Selected Titles in This Series

204 **K. Nomizu, Editor,** Selected papers on classical analysis

203 **N. N. Uraltseva, Editor,** Proceedings of the St. Petersburg Mathematical Society, Volume VII

202 **V. Turaev and A. Vershik, Editors,** Topology, ergodic theory, real algebraic geometry. Rokhlin's memorial

201 **Michael Semenov-Tian-Shansky, Editor,** L. D. Faddeev's seminar on mathematical physics

200 **L. Lerman, G. Polotovskiĭ, and L. Shilnikov, Editors,** Methods of qualitative theory of differential equations and related topics

199 **N. N. Uraltseva, Editor,** Proceedings of the St. Petersburg Mathematical Society, Volume VI

198 **R. A. Minlos, Senya Shlosman, and Yu. M. Suhov, Editors,** On Dobrushin's way. From probability theory to statistical physics

197 **Vladimir Arnold, Maxim Kontsevich, and Anton Zorich, Editors,** Pseudoperiodic topology

196 **Ya. Eliashberg, D. Fuchs, T. Ratiu, and A. Weinstein, Editors,** Northern California symplectic geometry seminar

195 **O. V. Belegradek et al.,** Model Theory and Applications

194 **A. Astashkevich and S. Tabachnikov, Editors,** Differential Topology, Infinite-Dimensional Lie Algebras, and Applications (D. B. Fuchs' 60th Anniversary Collection)

193 **N. N. Uraltseva, Editor,** Proceedings of the St. Petersburg Mathematical Society, Volume V

192 **Lev Beklemishev, Mati Pentus, and Nikolai Vereshchagin,** Provability, Complexity, Grammars

191 **A. Yu. Morozov and M. A. Olshanetsky, Editors,** Moscow Seminar in Mathematical Physics

190 **S. Tabachnikov, Editor,** Differential and Symplectic Topology of Knots and Curves

189 **V. Buslaev, M. Solomyak, and D. Yafaev, Editors,** Differential Operators and Spectral Theory (M. Sh. Birman's 70th anniversary collection)

188 **O. A. Ladyzhenskaya, Editor,** Proceedings of the St. Petersburg Mathematical Society, Volume IV

187 **M. V. Karasev, Editor,** Coherent Transform, Quantization, and Poisson Geometry

186 **A. Khovanskiĭ, A. Varchenko, and V. Vassiliev, Editors,** Geometry of Differential Equations

185 **B. Feigin and V. Vassiliev, Editors,** Topics in Quantum Groups and Finite-Type Invariants (Mathematics at the Independent University of Moscow)

184 **Peter Kuchment and Vladimir Lin, Editors,** Voronezh Winter Mathematical Schools (Dedicated to Selim Krein)

183 **K. Nomizu, Editor,** Selected Papers on Harmonic Analysis, Groups, and Invariants

182 **V. E. Zakharov, Editor,** Nonlinear Waves and Weak Turbulence

181 **G. I. Olshanski, Editor,** Kirillov's Seminar on Representation Theory

180 **A. Khovanskiĭ, A. Varchenko, and V. Vassiliev, Editors,** Topics in Singularity Theory

179 **V. M. Buchstaber and S. P. Novikov, Editors,** Solitons, Geometry, and Topology: On the Crossroad

178 **V. Kreinovich and G. Mints, Editors,** Problems of Reducing the Exhaustive Search

177 **R. L. Dobrushin, R. A. Minlos, M. A. Shubin, and A. M. Vershik, Editors,** Topics in Statistical and Theoretical Physics (F. A. Berezin Memorial Volume)

176 **E. V. Shikin, Editor,** Some Questions of Differential Geometry in the Large

(Continued in the back of this publication)

Selected Papers on Classical Analysis

American Mathematical Society
TRANSLATIONS
Series 2 • Volume 204

Selected Papers on Classical Analysis

Katsumi Nomizu
Editor

American Mathematical Society
Providence, Rhode Island

Translation edited by KATSUMI NOMIZU

2000 *Mathematics Subject Classification.* Primary 46J10, 34M55, 35K57, 35Q30, 57S30, 46Lxx.

Library of Congress Cataloging-in-Publication Data

Selected papers on classical analysis / Katsumi Nomizu, editor.
p. cm. — (American Mathematical Society translations, ISSN 0065-9290 ; ser. 2, v. 204)
Includes bibliographical references.
ISBN 0-8218-2780-4 (acid-free paper)
1. Mathematical analysis. I. Nomizu, Katsumi, 1924– II. Series.

QA3 .A572 ser. 2, vol. 204
[QA300.5]
510 s—dc21 2001022373
[515]

∞ The paper used in this book is acid-free and falls within the guidelines
established to ensure permanence and durability.
Visit the AMS home page at URL: http://www.ams.org/

10 9 8 7 6 5 4 3 2 1 03 02 01

Contents

Preface

This is a collection of several papers that originally appeared in the journal *Sūgaku* in Japanese. These translated articles would normally appear in the AMS journal *Sugaku Expositions*. In order to expedite publication, the AMS has chosen, with the consent of the Mathematical Society of Japan, to publish them as a volume of selected papers in the *Translations Series* 2 of the AMS.

The volume contains papers in the general area of mathematical analysis as it pertains to Free Probability, built based on operator algebras, in **Hiai**'s paper, and in **Matsuzaki**'s paper, to dynamics of Kleinian groups that originally goes back to the classical work by Fricke, Klein, and Poincaré and continues to Ahlfors and Bers, and develops into the recent work by Sullivan.

Nii's paper discusses stability analysis of travelling waves, and **Tanaka**'s article deals with extension of almost periodic functions and analyticity on related flows. One problem proposed by Forelli in 1970 on whether the function ring introduced by a minimal flow is a Dirichlet algebra is negatively answered.

Umemura's article discusses the Painlevé equations (known as $P_I, P_{II}, P_{III}, P_{IV}, P_V, P_{VI}$) that Painlevé discovered around 1890. Regarding whether they are irreducible and are really special functions, Poincaré had serious doubt. The last question was settled for P_I only in 1988. The validity in the general case is now convincing thanks to results related to representation theory—about 100 years after the original problem was posed.

Finally, **Yamazaki**'s article is summarized as follows: As an example of the close relationship between real analysis and nonlinear partial differential equations, the author surveys the recent work on the Navier-Stokes equations employing various function spaces, in particular, the work by Miyakawa on the decay of weak solutions, and that of Kozono and Yamazaki on the Cauchy problem with distributions as initial data.

Katsumi Nomizu

Amer. Math. Soc. Transl.
(2) Vol. **204**, 2001

Free Products of Operator Algebras and Free Probability Theory

Fumio Hiai

Introduction

A new probability theory called *free probability theory* has been extensively developed since Dan Voiculescu discovered it in the early 1980s. This is a sort of noncommutative probability theory closely related with free products of operator algebras, and the notion of freeness plays a central role there, just as independence does in classical probability theory. Very briefly speaking,

"Free probability theory = noncommutative probability theory + freeness."

Noncommutative probability spaces are defined algebraically; in general, a pair $(\mathcal{A}, \varphi)$ of a $(*\text{-})$algebra $\mathcal{A}$ with unit $\mathbf{1}$ and a linear functional φ with $\varphi(\mathbf{1}) = 1$ (also $\varphi(a^*) = \overline{\varphi(a)}$ if $\mathcal{A}$ is a $*$-algebra) is called a *noncommutative probability space*. A C^*-algebra or von Neumann algebra is sometimes taken as $\mathcal{A}$; in this case, φ is assumed to be a state on $\mathcal{A}$ (a faithful normal state if $\mathcal{A}$ is a von Neumann algebra) and $(\mathcal{A}, \varphi)$ is called a *C^*-probability space* or a *W^*-probability space*. In particular, the case of a von Neumann algebra with a faithful normal tracial state is important, and it is called a *tracial W^*-probability space*. A classical probability space $(\Omega, \mathcal{F}, \mathbf{P})$ is reformulated in the setting of a noncommutative probability space: $\mathcal{A} = L^\infty(\Omega)$ and $\varphi = E(\cdot) = \int \cdot \, d\mathbf{P}$.

First of all, let us briefly recall some basic terminology about operator algebras. An algebra is a complex vector space equipped with a product so that both left and right multiplications are linear mappings. A $*$-algebra is an algebra with an involution satisfying $(\alpha a + \beta b)^* = \bar{\alpha} a^* + \bar{\beta} b^*$, $(ab)^* = b^* a^*$ and $(a^*)^* = a$. A C^*-algebra is a $*$-algebra that is a Banach space with a norm satisfying $\|a^*\| = \|a\|$ and $\|a^* a\| = \|a\|^2$. Every C^*-algebra is realized as a closed (in the operator norm) $*$-subalgebra of $B(\mathcal{H})$, the algebra of all bounded linear operators on a Hilbert space $\mathcal{H}$. Moreover, a von Neumann algebra is a $*$-subalgebra of $B(\mathcal{H})$ that is closed in the weak operator topology. A linear functional $\varphi : \mathcal{A} \to \mathbb{C}$ on a C^*-algebra $\mathcal{A}$ (with $\mathbf{1}$) is called a state on $\mathcal{A}$ if $\varphi(a^* a) \ge 0$ and $\varphi(\mathbf{1}) = 1$. A state φ on a von Neumann algebra $\mathcal{M}$ is normal if $0 \le a_\alpha \nearrow a$ implies $\varphi(a_\alpha) \nearrow \varphi(a)$. An abelian C^*-algebra (with $\mathbf{1}$) is isometrically isomorphic to $C(\Omega)$, the algebra of continuous functions

---

2000 *Mathematics Subject Classification.* Primary 46L54.
This article originally appeared in Japanese in Sūgaku **51** (4) (1999), 377–394.

classical probability	free probability
probability space $(\Omega, \mathcal{F}, P)$	noncommutative probability space $(\mathcal{A}, \varphi)$
random variable X	noncommutative random variable $a \in \mathcal{A}$
expectation $E(X) = \int X\,dP$	expectation $\varphi(a)$
independence	freeness
tensor product	free product
central limit theorem	free central limit theorem
Gauss (normal) law	Wigner (semicircle) law
Boltzmann-Gibbs entropy	free entropy

on a compact Hausdorff space Ω. When $\mathcal{M}$ is an abelian von Neumann algebra and φ is a faithful normal state on $\mathcal{M}$, $(\mathcal{M}, \varphi)$ is represented as $(L^\infty(\Omega), E)$ for some probability space $(\Omega, \mathcal{F}, \mathbf{P})$. In this way, noncommutative probability theory on C^*-algebras or von Neumann algebras is a generalization of classical probability theory, including the classical theory as a particular case of abelian algebras.

Free probability theory and classical theory have surprising parallelisms in their frames while they are very different in the details. Some parallelisms between them are given in the table. The free product of operator algebras or the operator algebra generated by a free family of noncommutative random variables bears quite high noncommutativity, and they become something like free group factors. From the viewpoint of operator algebras, free probability theory provides techniques for analyzing free group factors and their relatives. Another remarkable point is that random matrices and free entropy are powerful tools in free probability theory. In the sequel we review some main streams in free probability. In Section 1, based on [111], some definitions of free products related with operator algebras are summarized; they are not so simple, and so the reader without a good background in operator algebras may start from Section 2.

1. Definitions of free products of operator algebras

Free products, as well as tensor products, are defined in various algebraic categories, but most fundamental is the free product of discrete groups. For discrete groups G_1 and G_2 (with respective units e_1, e_2), take the free group (the set of all words) generated by all elements of $(G_1 \setminus \{e_1\}) \sqcup (G_2 \setminus \{e_2\})$ and divide it by all original relations of elements in G_1 or in G_2; then we have the free product group $G_1 * G_2$. For example, the n-fold free product $\mathbb{Z}_1 * \cdots * \mathbb{Z}$ of $\mathbb{Z}$ is the free group $\mathbf{F}_n$. It is well known that if G_1, G_2 are not trivial in the sense that $G_i \neq \{e_i\}$, then $G_1 * G_2$ is necessarily non-amenable except in the case $G_1 = G_2 = \mathbb{Z}_2$. In this way, $G_1 * G_2$ becomes a group of high degree of noncommutativity even when G_1 and G_2 are abelian.

The free product of $*$-algebras is defined in a fashion similar to the group case. We first introduce the free product of two noncommutative probability spaces $(\mathcal{A}_1, \varphi_1)$ and $(\mathcal{A}_2, \varphi_2)$. Set $\mathcal{A}_i^0 := \{a \in \mathcal{A}_i : \varphi_i(a) = 0\}$ (so $\mathcal{A}_i = \mathbb{C}\mathbf{1} \oplus \mathcal{A}_i^0$) and define

$$\mathcal{A} := \mathbb{C}\mathbf{1} \oplus \bigoplus_{n=1}^{\infty} \left(\bigoplus_{i_1 \neq i_2 \neq \cdots \neq i_n} \mathcal{A}_{i_1}^0 \otimes \mathcal{A}_{i_2}^0 \otimes \cdots \otimes \mathcal{A}_{i_n}^0 \right),$$

which naturally becomes a $*$-algebra. Moreover, define a linear functional φ on $\mathcal{A}$ by $\varphi : \lambda\mathbf{1} + \sum a_1 a_2 \cdots a_n \mapsto \lambda$ (the coefficient of $\mathbf{1}$); then $(\mathcal{A}, \varphi)$ is a noncommutative probability space. This is called the *free product* of $(\mathcal{A}_1, \varphi_1)$ and $(\mathcal{A}_2, \varphi_2)$, and we write $(\mathcal{A}, \varphi) = (\mathcal{A}_1, \varphi_1) * (\mathcal{A}_2, \varphi_2) = (\mathcal{A}_1 * \mathcal{A}_2, \varphi_1 * \varphi_2)$. For an arbitrary number of noncommutative probability spaces $(\mathcal{A}_i, \varphi_i)$ $(i \in I)$ one can similarly define $\ast_{i\in I}(\mathcal{A}_i, \varphi_i) = (\ast_{i\in I}\mathcal{A}_i, \ast_{i\in I}\varphi_i)$.

When $\mathcal{A}_i$ are C^*-algebras (resp. von Neumann algebras) in the above, a certain completion procedure must be done to make the free product $\ast_{i\in I}\mathcal{A}_i$ a C^*-algebra (resp. a von Neumann algebra). When $\mathcal{A}_i$ are C^*-algebras, the universal C^*-completion of the $*$-algebra $\ast_{i\in I}\mathcal{A}_i$ defined above is called the *universal free product C^*-algebra* and is denoted by $\ast_{i\in I}\mathcal{A}_i$ again. On the other hand, to define the (reduced) C^*-free product of C^*-probability spaces $(\mathcal{A}_i, \varphi_i)$ $(i \in I)$, it is convenient to first introduce the free product of representations of $*$-algebras. To do so, we start from the free product of Hilbert spaces. For a family of Hilbert spaces $(\mathcal{H}_i, \xi_i)$ $(i \in I)$ with distinguished unit vectors ξ_i, set $\mathcal{H}_i^0 := \mathcal{H}_i \ominus \mathbb{C}\xi_i$ and define the *free product Hilbert space* $(\mathcal{H}, \xi) = \ast_{i\in I}(\mathcal{H}_i, \xi_i)$ (ξ is a unit vector) by

$$\mathcal{H} := \mathbb{C}\xi \oplus \bigoplus_{n=1}^{\infty}\left(\bigoplus_{i_1\neq i_2\neq\cdots\neq i_n} \mathcal{H}_{i_1}^0 \otimes \mathcal{H}_{i_2}^0 \otimes \cdots \otimes \mathcal{H}_{i_n}^0\right).$$

The linear structure and the inner product in $\mathcal{H}$ are defined similarly to the construction of Fock spaces, and ξ can be considered as the vacuum vector. For each $i \in I$, we get a subspace

$$\mathcal{H}(i) := \mathbb{C}\xi \oplus \bigoplus_{n=1}^{\infty}\left(\bigoplus_{i\neq i_1\neq i_2\neq\cdots\neq i_n} \mathcal{H}_{i_1}^0 \otimes \mathcal{H}_{i_2}^0 \otimes \cdots \otimes \mathcal{H}_{i_n}^0\right)$$

and define a unitary operator $V_i : \mathcal{H}_i \otimes \mathcal{H}(i) \to \mathcal{H}$ as follows:

$$V_i : \begin{cases} \xi_i \otimes \xi \mapsto \xi, \\ \eta \otimes \xi \mapsto \eta & (\eta \in \mathcal{H}_i^0), \\ \xi \otimes \zeta \mapsto \zeta & (\zeta \in \mathcal{H}_{i_1}^0 \otimes \cdots \otimes \mathcal{H}_{i_n}^0), \\ \eta \otimes \zeta \mapsto \eta \otimes \zeta & (\eta \in \mathcal{H}_i^0,\ \zeta \in \mathcal{H}_{i_1}^0 \otimes \cdots \otimes \mathcal{H}_{i_n}^0). \end{cases}$$

For each $i \in I$, let π_i be a representation of a $*$-algebra $\mathcal{A}_i$ on a Hilbert space $(\mathcal{H}_i, \xi_i)$. Using the representations $\lambda_i : \mathcal{A}_i \to B(\mathcal{H})$ $(i \in I)$ given by

$$\lambda_i(a) := V_i(\pi_i(a) \otimes \mathbf{1}_{\mathcal{H}(i)})V_i^* \qquad (a \in \mathcal{A}_i),$$

one can define a representation π of $\ast_{i\in I}\mathcal{A}_i$ as

$$\pi(a_1 a_2 \cdots a_n) := \lambda_{i_1}(a_1)\lambda_{i_2}(a_2)\cdots\lambda_{i_n}(a_n)$$

for $a_k \in \mathcal{A}_{i_k}$ $(1 \leq k \leq n)$, $i_1 \neq i_2 \neq \cdots \neq i_n$, which is called the *free product representation* of π_i and is denoted by $\pi = \ast_{i\in I}\pi_i$. Given C^*-probability spaces $(\mathcal{A}_i, \varphi_i)$ let $\{\pi_i, \mathcal{H}_i, \xi_i\}$ be the GNS cyclic representations, i.e., $\varphi_i(a) = \langle \pi_i(a)\xi_i, \xi_i\rangle$ $(a \in \mathcal{A}_i)$, $\mathcal{H}_i = \overline{\pi_i(\mathcal{A}_i)\xi_i}$. Then one has the free product representation $\pi := \ast_{i\in I}\pi_i$ on $(\mathcal{H}, \xi) := \ast_{i\in I}(\mathcal{H}_i, \xi_i)$ and hence the state $\varphi(a) := \langle \pi(a)\xi, \xi\rangle$ on the universal free product C^*-algebra $\mathcal{A} := \ast_{i\in I}\mathcal{A}_i$. This φ is called the *free product state* of φ_i and is denoted by $\ast_{i\in I}\varphi_i$. Here it is easy to see that $\{\pi, \mathcal{H}, \xi\}$ is the GNS cyclic representation of $(\ast_{i\in I}\mathcal{A}_i, \ast_{i\in I}\varphi_i)$. The *(reduced) free product C^*-algebra* of $(\mathcal{A}_i, \varphi_i)$ is defined as $(\pi(\ast_{i\in I}\mathcal{A}_i), \langle\cdot\,\xi, \xi\rangle)$ and is denoted by $\ast_{i\in I}(\mathcal{A}_i, \varphi_i)$ or again $(\ast_{i\in I}\mathcal{A}_i, \ast_{i\in I}\varphi_i)$.

Finally, we define the free product of von Neumann algebras. Let $(\mathcal{M}_i, \varphi_i)$ $(i \in I)$ be W^*-probability spaces, that is, $\mathcal{M}_i$ is a von Neumann algebra with a faithful normal state φ_i. $\mathcal{M}_i$ is faithfully represented on a Hilbert space $(\mathcal{H}_i, \xi_i)$ with a cyclic and separating vector ξ_i by the GNS representation associated with φ_i. Fix the representations $\lambda_i(a) := V_i(a \otimes \mathbf{1}_{\mathcal{H}(i)})V_i^*$ $(a \in \mathcal{M}_i)$ of $\mathcal{M}_i$ on $(\mathcal{H}, \xi) := \ast_{i\in I}(\mathcal{H}_i, \xi_i)$ and define the *free product von Neumann algebra*

$$(\mathcal{M}, \varphi) = \ast_{i\in I}(\mathcal{M}_i, \varphi_i) = (\ast_{i\in I}\mathcal{M}_i, \ast_{i\in I}\varphi_i)$$

by

$$\mathcal{M} := \left(\bigcup_{i\in I} \lambda_i(\mathcal{M}_i)\right)'', \quad \varphi := \langle \cdot\, \xi, \xi \rangle .$$

Note that ξ is a cyclic and separating vector for $\mathcal{M}$.

EXAMPLE. For a discrete group G, the left and right regular representations on the Hilbert space $\ell^2(G)$ are given by

$$(L_g\xi)(h) := \xi(g^{-1}h), \quad (R_g\xi)(h) := \xi(hg) \qquad (\xi \in \ell^2(G),\ g, h \in G).$$

The *reduced group C^*-algebra* of G is $C_r^*(G) := \overline{\mathrm{span}}^{\|\cdot\|}\{L_g : g \in G\}$ generated by L_g's, and the *group von Neumann algebra* of G is $\mathcal{L}(G) := \{L_g : g \in G\}''$. $\delta_e \in \ell^2(G)$ is a cyclic and separating trace vector for $\mathcal{L}(G)$, and one has $\mathcal{L}(G)' = R(G) := \{R_g : g \in G\}''$. When G is an infinite group, $\mathcal{L}(G)$ is a type II_1 factor if and only if G is an ICC group (i.e., the conjugacy class of any $g \in G\setminus\{e\}$ is infinite). When $G := \ast_{i\in I}G_i$ is the free product group of discrete groups G_i $(i \in I)$, one can easily see that

$$(\ell^2(G), \delta_e) = \ast_{i\in I}(\ell^2(G_i), \delta_{e_i}),$$

$$C_r^*(G) = \ast_{i\in I}C_r^*(G_i) \quad \text{(the reduced free product } C^*\text{-algebra)},$$

$$\mathcal{L}(G) = \ast_{i\in I}\mathcal{L}(G_i) \quad \text{(the free product von Neumann algebra)}.$$

In particular, since $\mathbf{F}_n = \ast_{i=1}^n \mathbb{Z}$, one has $\mathcal{L}(\mathbf{F}_n) = \ast_{i=1}^n \mathcal{L}(\mathbb{Z})$, that is, the *free group factor* $\mathcal{L}(\mathbf{F}_n)$ is represented as the n-fold free product of the commutative von Neumann algebra $\mathcal{L}(\mathbb{Z})$ $(\cong L^\infty[0,1])$.

To conclude this section, we mention three preceding papers on the (amalgamated) free product of operator algebras; recent developments will be discussed in Section 5. The free product of operator algebras was first treated in [22], where the free product of von Neumann algebras with cyclic and separating trace vectors was introduced and the following result was given: If $\mathcal{M}_1$ and $\mathcal{M}_2$ are von Neumann algebras with faithful normal tracial states such that $\dim \mathcal{M}_1 \ge 2$ and $\dim \mathcal{M}_2 \ge 3$, and if $\mathcal{M}_1$ and $\mathcal{M}_2$ have orthogonal unitary bases, then $\mathcal{M}_1 * \mathcal{M}_2$ is a type II_1 factor and it does not have *property* Γ. Murray and von Neumann formerly proved that the hyperfinite type II_1 factor has property Γ but $\mathcal{L}(\mathbf{F}_2)$ does not (so that $\mathcal{L}(\mathbf{F}_2)$ is the first example of a non-hyperfinite factor). On the other hand, the free product C^*-algebra (associated with states) was introduced in [4, 96]. Voiculescu's paper [96] contains many fundamental ideas, including the freeness for instance, and thus providing the foundation of the recent breakthrough of free probability theory. Avitzour's paper [4] actually reached the concept of freeness too; however, he just recognized it as a technique for defining the free product of states. But a sufficient condition given in [4] for the simplicity of free product C^*-algebras (or the factoriality in the case of free product von Neumann algebras) is important and useful.

2. Basics of free probability theory

Let $(\mathcal{A}, \varphi)$ be a noncommutative probability space and $a \in \mathcal{A}$ a noncommutative random variable. The kth moment of a is $\varphi(a^k)$ ($k \in \mathbb{N}$). The linear extension of the moment sequence to the polynomial algebra $\mathbb{C}\langle X\rangle$ given by $\mu(P(X)) := \varphi(P(a))$ ($P \in \mathbb{C}\langle X\rangle$) is called the *distribution* of a. In particular, when $\mathcal{A}$ is a C^*-algebra and a is a normal element, the distribution of a can be realized as a probability measure supported on the spectrum $\sigma(a)$ so that $\varphi(f(a)) = \int_{\sigma(a)} f(x)\,d\mu(x)$ for $f \in C(\sigma(a))$. A sequence (a_n) of noncommutative random variables is said to *converge in distribution* or *converge in moments* if the kth moment of a_n converges to that of a for each k.

First, let us state the definition of freeness that is fundamental in free probability theory. In a noncommutative probability space $(\mathcal{A}, \varphi)$, a family $(\mathcal{A}_i)_{i\in I}$ of subalgebras (containing $\mathbf{1}$) of $\mathcal{A}$ is said to be *free* if

$$a_k \in \mathcal{A}_{i_k},\ \varphi(a_k) = 0\ (1 \le k \le n) \text{ implies } \varphi(a_1 a_2 \cdots a_n) = 0$$

for every $i_1, \dots, i_n \in I$, $i_1 \ne i_2 \ne \cdots \ne i_n$ (non-neighboring i_k's are allowed to be equal). This definition can be viewed as a version of independence in classical probability theory. Families of independent (commutative) random variables can be constructed by tensor products, while families of free (noncommutative) random variables can be realized by free products. Indeed, let $(\mathcal{A}, \varphi) = \ast_{i\in I}(\mathcal{A}_i, \varphi_i)$ be a free product noncommutative probability space and let $j_i : \mathcal{A}_i \hookrightarrow \mathcal{A}$ be the natural embedding. Then it is clear from the definition of free products that $(j_i(\mathcal{A}_i))_{i\in I}$ is free, so the definition of freeness is just a restatement of the characteristic of free products. When $\mathcal{X}_i$ ($i \in I$) are subsets of $\mathcal{A}$, $(\mathcal{X}_i)_{i\in I}$ is said to be *free* if the subalgebras $\mathrm{Alg}(\mathcal{X}_i \cup \{\mathbf{1}\})$ generated by $\mathcal{X}_i \cup \{\mathbf{1}\}$ ($i \in I$) are free. Also it is $*$-*free* if $\mathcal{A}$ is a $*$-algebra and the $*$-subalgebras generated by $\mathcal{X}_i \cup \{\mathbf{1}\}$ ($i \in I$) are free.

EXAMPLE. Consider a noncommutative probability space $(B(\mathcal{F}(\mathcal{H})), \varphi)$, where $\mathcal{F}(\mathcal{H}) := \mathbb{C}\Phi \oplus \bigoplus_{n=1}^{\infty} \mathcal{H}^{\otimes n}$ is the *full Fock space* over a Hilbert space $\mathcal{H}$ (Φ being a vacuum vector) and φ is the vacuum expectation. The (*left*) *creation operator* for a vector $h \in \mathcal{H}$ is defined by

$$\ell(h)\Phi := h\,, \quad \ell(h)(\eta_1 \otimes \cdots \otimes \eta_n) := h \otimes \eta_1 \otimes \cdots \otimes \eta_n\,.$$

The *annihilation operator* $\ell(h)^*$ is given as

$$\ell(h)^*\Phi = 0\,, \quad \ell(h)^*(\eta_1 \otimes \cdots \otimes \eta_n) = \langle \eta_1, h\rangle \eta_2 \otimes \cdots \otimes \eta_n\,.$$

If $\|h\| = 1$, then $\ell(h)^*\ell(h) = \mathbf{1}$ and the selfadjoint $s(h) := \ell(h)^* + \ell(h)$ has the *Wigner* (*semicircle*) *distribution*

$$w_2 := \frac{1}{2\pi}\sqrt{4 - x^2}\,\chi_{[-2,2]}(x)\,dx$$

of radius 2. The moments of w_2 are $m_{2k-1}(w_2) = 0$ and $m_{2k}(w_2) = \binom{2k}{k}/(k+1)$. This $2k$th moment is the famous *Catalan number.* When $(h_i)_{i\in I}$ is an orthonormal set in $\mathcal{H}$, $(\ell(h_i))_{i\in I}$ is $*$-free and so $(s(h_i))_{i\in I}$ is free. Concerning the C^*-algebra $C^*(\ell(\mathcal{H}))$ generated by $\{\ell(h) : h \in \mathcal{H}\}$, it is known that $C^*(\ell(\mathcal{H})) \supset \mathcal{K}$ and $C^*(\ell(\mathcal{H}))/\mathcal{K} \cong \mathcal{O}_n$ if $\dim \mathcal{H} = n < \infty$, and $C^*(\ell(\mathcal{H})) \cong \mathcal{O}_\infty$ if $\dim \mathcal{H} = \infty$, where $\mathcal{K}$ is the ideal of compact operators and $\mathcal{O}_n$ is the so-called Cuntz algebra. Also, $\{s(h) : h \in \mathcal{H}\}$ generates the free group factor: $s(\mathcal{H})'' \cong \mathcal{L}(\mathbf{F}_{\dim \mathcal{H}})$.

Fock spaces useful in the description of quantum statistics are symmetric (Bose) Fock spaces and anti-symmetric (Fermi) Fock spaces. The creation and annihilation operators a^*, a on a symmetric Fock space satisfy the commutation relation $aa^* - a^*a = \mathbf{1}$, and it is well known that the distribution of $a+a^*$ at the vacuum vector is the standard normal distribution $N(0,1)$. On the other hand, on an anti-symmetric Fock space the commutation relation is $aa^* + a^*a = \mathbf{1}$, and the distribution of $a+a^*$ is $\frac{1}{2}(\delta_{-1} + \delta_1)$. Interpolating these commutation relations, one can consider the *q-commutation relation* for $q \in [-1, 1]$ as follows:

$$aa^* - qa^*a = \mathbf{1} \quad (\text{also } a_i a_j^* - q a_j^* a_i = \delta_{ij}\mathbf{1},\ 1 \le i, j \le n)$$

($q = 1, 0, -1$ are the symmetric, free, anti-symmetric cases, respectively). The distribution of this $a + a^*$ is called the *q-Gaussian distribution.* See [18, 19, 35, 58, 63, 81] for q-commutation relations.

The central limit theorem in classical probability theory says that if (X_n) is an independent sequence of random variables that are identically distributed (or satisfy a certain milder assumption), then $(X_1 + \cdots + X_n)/\sqrt{n}$ converges in law to a normal distribution. Early in [96] Voiculescu gave a free probabilistic analogue of this. The semicircle (Wigner) law appears as the limit distribution of this *free central limit theorem*, which justifies the statement that the semicircle distribution is the free analogue of the normal one.

THEOREM ([96]). *If a sequence (a_n) is free in a noncommutative probability space $(\mathcal{A}, \varphi)$ and if $\varphi(a_n) = 0$, $\lim_{n\to\infty} \frac{1}{n} \sum_{i=1}^n \varphi(a_i^2) = 1$ and $\sup_i |\varphi(a_i^k)| < +\infty$ $(k \in \mathbb{N})$, then $(a_1 + \cdots + a_n)/\sqrt{n}$ converges in distribution to the semicircle law w_2.*

In classical theory, when the distributions of independent X_1 and X_2 are μ_1 and μ_2, respectively, the distribution of $X_1 + X_2$ is the convolution $\mu_1 * \mu_2$. A free probabilistic analogue of this can be introduced. When μ_1 and μ_2 are the distributions (as moment sequences) of free noncommutative random variables a_1 and a_2, respectively, define the *additive free convolution* $\mu_1 \boxplus \mu_2$ as the distribution of $a_1 + a_2$, and the *multiplicative free convolution* $\mu_1 \boxtimes \mu_2$ as the distribution of $a_1 a_2$. The free convolutions $\mu_1 \boxplus \mu_2$ and $\mu_1 \boxtimes \mu_2$ are determined only by the distributions μ_1, μ_2, independently of the choice of a_1 and a_2 as well as $(\mathcal{A}, \varphi)$; in fact, such a_1 and a_2 can be chosen by taking a free product.

The R-transform (R-series), an analogue of the logarithm of the Fourier transform, is useful for computing the additive free convolution explicitly. Let $\mathcal{E}$ denote the set of all formal infinite series $T = \sum_{m,n=0}^\infty c_{mn} \ell^m \ell^{*n}$ such that $c_{mn} = 0$ $(n > N)$ for some $N \in \mathbb{N}$. $\mathcal{E}$ is a linear space by natural operations, and the multiplication is defined in $\mathcal{E}$ under the single relation $\ell^*\ell = \mathbf{1}$ (multiply T, T' term by term, reduce by $\ell^*\ell = \mathbf{1}$ and sum the coefficients of the same $\ell^m \ell^{*n}$). Furthermore, define $\omega(T) := c_{00}$, the constant term of T. Then $(\mathcal{E}, \omega)$ becomes a noncommutative probability space (without $*$-operation). For every noncommutative random variable a there exists a unique

$$T = \ell^* + \sum_{n=0}^{\infty} \alpha_{n+1}\ell^n \in \mathcal{E}$$

such that $\varphi(a^k) = \omega(T^k)$ $(k \in \mathbb{N})$. In fact, one can define $\alpha_1, \alpha_2, \ldots$ recursively. Then the *R-transform* of a or (μ, the distribution of a) is defined by $R_a(z) = R_\mu(z) := \sum_{n=0}^{\infty} \alpha_{n+1} z^n$. (Sometimes $R_a(z) = \sum_{n=1}^{\infty} \alpha_n z^n$, different by a factor z, is utilized.) Also, the coefficients α_n of the R-series are called the *free cumulants* of a (or μ).

The Cauchy transform of a probability measure μ on $\mathbb{R}$ is

$$G_\mu(z) := \int_{\mathbb{R}} (z - x)^{-1}\, d\mu(x),$$

which is formally expanded as $G_\mu(z) = \sum_{k=0}^{\infty} (\int x^k\, d\mu(x)) z^{-k-1}$. So, for a general distribution μ (not necessarily a measure), we define the formal series

$$G_\mu(z) := z^{-1} + \sum_{k=1}^{\infty} \mu(X^k) z^{-k-1}\,.$$

THEOREM ([97]; also [86]). *For each distribution μ, $G_\mu(z)$ and $K_\mu(z) := z^{-1} + R_\mu(z)$ are the converse of each other as formal power series. Moreover,*

$$R_{\mu_1 \boxplus \mu_2}(z) = R_{\mu_1}(z) + R_{\mu_2}(z)$$

for all distributions μ_1, μ_2.

In this way, the free convolution $\mu_1 \boxplus \mu_2$ is linearized by the R-transform. To determine $\mu = \mu_1 \boxplus \mu_2$ for compactly supported probability measures μ_1, μ_2 on $\mathbb{R}$, proceed as $\mu_i \to G_{\mu_i} \to K_{\mu_i} \to R_{\mu_i}$ and then go back as $R_\mu = R_{\mu_1} + R_{\mu_2} \to K_\mu \to G_\mu \to \mu$. The Stieltjes inversion formula is useful for the last $G_\mu \to \mu$. The notion of R-series was also discovered in a particular situation from the study of random walks on free product groups (see, e.g., [21, 114]). A similar device, called an *S-transform*, is employed for the multiplicative free convolution ([98], also [45]).

The semicircle distribution $w_{2\sqrt{t}}$ of mean 0 and radius $2\sqrt{t}$ $(t \geq 0)$ forms a semigroup with respect to $\boxplus$: $w_{2\sqrt{t_1}} \boxplus w_{2\sqrt{t_2}} = w_{2\sqrt{t_1+t_2}}$. If μ_t $(t \geq 0)$ and ν are compactly supported probability measures on $\mathbb{R}$ and $(\mu_t)_{t\geq 0}$ is a $\boxplus$-semigroup, then the Cauchy transform $G(z,t)$ of $\nu \boxplus \mu_t$ satisfies the complex Burger equation

$$\frac{\partial G(z,t)}{\partial t} + R_{\mu_1}(G(z,t)) \frac{\partial G(z,t)}{\partial z} = 0\,,$$

which plays an important role in free probability theory. In particular, when $\mu_t = w_{2\sqrt{t}}$, $\frac{\partial G}{\partial t} + G \frac{\partial G}{\partial z} = 0$ is considered as a free analogue of the heat equation.

The following are typical examples of noncommutative random variables in C^*-probability spaces:

1. A *semicircular element* is a selfadjoint element having the semicircle distribution w_2.
2. A *quarter-circular element* is a positive element having the quarter-circular distribution $\frac{1}{\pi}\sqrt{4 - x^2}\chi_{[0,2]}(x)\, dx$.
3. A *Haar unitary* is a unitary whose distribution is the uniform distribution on the unit circle $\mathbb{T}$.

4. A *circular element* is written as $c = (a + \mathrm{i}b)/\sqrt{2}$ with free semicircular elements a, b. $c = v\,|c|$ in the polar decomposition is circular if and only if v is a Haar unitary, $|c|$ is quarter-circular and $v, |c|$ are free.
5. A *free Poisson element* is a positive element whose R-series is $\lambda z/(1-z)$ for some $\lambda > 0$ (i.e., all free cumulants are λ). The distribution of this is called the *free Poisson distribution*, and it coincides with the limit distribution in the free Poisson limit theorem ([111]). The square of a semicircular element or a quarter-circular element is a free Poisson element with $\lambda = 1$.

A free family of semicircular elements is called a *semicircular system*. When noncommutative random variables $a_1, \dots, a_N$ are free, some characterizations of $(a_i)_{1 \le i \le N}$ being a semicircular system are known ([64, 54, 55]). Haagerup [44] proved, using semicircular systems, that quasitraces on exact C^*-algebras are traces. This result was recently reproved in [47] by using random matrices.

The notion of R-series was generalized in [64] (also [86]) to the multivariate R-series $R_{(a_1,\dots,a_N)}(X_1,\dots,X_N)$ for an N-tuple $(a_1,\dots,a_N)$ of noncommutative random variables, which is a formal power series of noncommuting indeterminates $X_1,\dots,X_N$.

THEOREM. [64] *In a noncommutative probability space $(\mathcal{A},\varphi)$, the following properties hold:*

(1) *Two tuples $(a_1,\dots,a_N)$ and $(b_1,\dots,b_L)$ are free if and only if*

$$R_{(a_1,\dots,a_N,b_1,\dots,b_L)}(X_1,\dots,X_N,X_{N+1},\dots,X_{N+L}) \\ = R_{(a_1,\dots,a_N)}(X_1,\dots,X_N) + R_{(b_1,\dots,b_L)}(X_{N+1},\dots,X_{N+L}).$$

(2) *If two N-tuples $(a_1,\dots,a_N)$ and $(b_1,\dots,b_N)$ are free, then*

$$R_{(a_1+b_1,\dots,a_N+b_N)} = R_{(a_1,\dots,a_N)}(X_1,\dots,X_N) + R_{(b_1,\dots,b_N)}(X_1,\dots,X_N).$$

A pair (a,b) of noncommutative random variables is said to be *R-diagonal* if the R-series of (a,b) is written in the following form:

$$R_{(a,b)}(X_1,X_2) = \sum_{n=1}^{\infty} \bigl[\alpha_n (X_1X_2)^n + \beta_n (X_2X_1)^n\bigr],$$

that is, all coefficients except those of alternating $X_1X_2\cdots X_1X_2$ or $X_2X_1\cdots X_2X_1$ vanish. Note that $\alpha_n = \beta_n$ must hold in $(\mathcal{A},\varphi)$ with φ tracial. Furthermore, when (a,a^*) is R-diagonal, a is called an *R-diagonal element*. For example, a circular element c and a Haar unitary u are R-diagonal:

$$R_{(c,c^*)}(X_1,X_2) = X_1X_2 + X_2X_1,$$

$$R_{(u,u^*)}(X_1,X_2) = \sum_{n=1}^{\infty} (-1)^n c_{n-1} \bigl[(X_1X_2)^n + (X_2X_1)^n\bigr],$$

where $c_n = \binom{2n}{n}/(n+1)$, the Catalan numbers. See [69, 67, 57] for R-diagonal elements.

Free probability theory is of course primarily based on operator algebras and probability theory, but we may stress that it also enjoys powerful devices from other disciplines such as combinatorics and complex analysis. A combinatorial method using non-crossing partitions was developed by Speicher [86, 87, 68–70]. A complex analysis method using the theory of Nevanlinna-Pick functions is useful in the study of infinitely divisible distributions with respect to free convolution. [10] is a refinement of the free central limit theorem using a method from complex analysis. Compound free Poisson distributions are $\boxplus$-infinitely divisible ones generalizing free Poisson distributions; any $\boxplus$-infinitely divisible distribution can be the limit of a sequence of compound free Poisson distributions ([53, 87]). P. Biane [13] showed that some asymptotic behavior of irreducible representations of $\mathbf{S}_N$ as $N \to \infty$ can be described in terms of the free cumulants of the distribution determined by the limiting form of corresponding Young diagrams. This is an interesting application of free probability to the group representation theory. Moreover, the study of free stochastic processes has been developed recently (e.g. [11, 12, 14, 15]). Orthogonal polynomials are sometimes useful in the description of distributions (e.g. [2, 17, 63]), and are closely related with random matrices treated in the next section.

3. Random matrices and asymptotic freeness

Random matrices are matrices whose entries are random variables. Let M_n denote the algebra of $n \times n$ complex matrices, M_n^{sa} the set of all selfadjoint matrices in M_n, and tr_n the normalized trace on M_n. Fix a classical probability space $(\Omega, \mathcal{F}, \mathbf{P})$ and let $\mathcal{L} := \bigcap_{1 \le p < \infty} L^p(\Omega, \mathbf{P})$ ($\supsetneqq L^\infty(\Omega, \mathbf{P})$), the space of all random variables all of whose moments are finite. Then $M_n(\mathcal{L}) = M_n \otimes \mathcal{L}$, the set of all $n \times n$ matrices whose entries are elements of $\mathcal{L}$, becomes a $*$-algebra. We fix the noncommutative probability space $(M_n(\mathcal{L}), \mathrm{tr}_n)$ with a tracial state $\tau_n(X) := \frac{1}{n}\sum_{i=1}^n E(X_{ii}) = E(\mathrm{tr}_n(X))$ as a working space of $n \times n$ random matrices.

When $\lambda_1(X), \dots, \lambda_n(X)$ are n (random) eigenvalues of $X \in M_n(\mathcal{L})$, the *empirical spectral density* (or *empirical eigenvalue distribution*) of X is defined by $R_X := \frac{1}{n}\sum_{i=1}^n \delta_{\lambda_i(X)}$ and the *mean spectral density* of X by $\mu_X := E(R_X)$, the expectation of R_X. R_X is a random (atomic) probability measure on $\mathbb{C}$ (on $\mathbb{R}$ if X is selfadjoint, on $\mathbb{T}$ if X is unitary), and μ_X is a probability measure on $\mathbb{C}$. The classical *Wigner limit theorem* is:

THEOREM ([112, 113]). *Let $X(n)$ be a real symmetric or complex selfadjoint $n \times n$ random matrix such that $X_{ij}(n)$ $(1 \le i \le j \le n)$ are independent. Assume that $E(X_{ij}(n)) = 0$, $E(|X_{ij}(n)|^2) = 1$ for $1 \le i < j \le n$, and further, for each $k \in \mathbb{N}$, $\{E(|X_{ij}(n)|^k) : 1 \le i \le j \le n, n \in \mathbb{N}\}$ is bounded. Then $\mu_{X(n)/\sqrt{n}}$ converges to the Wigner distribution w_2. (Indeed, the stronger almost sure convergence is known to hold* [3]: *$R_{X(n)/\sqrt{n}}$ converges a.s. to w_2.)*

The above limit theorem and the free central limit theorem in the previous section have the same limit distribution w_2. This fact suggests that there should be some connection between random matrices and freeness, and it was actually discovered by Voiculescu [100] as the asymptotic freeness of random matrices. The asymptotic freeness means the approach to freeness in the limit; the precise definition is given as follows.

Now let $(\mathcal{A}_n, \varphi_n)$ $(n \in \mathbb{N})$ be noncommutative probability spaces (for example, $(M_n(\mathcal{L}), \tau_n)$ for the case of random matrices). Let S be a set and $S = \bigcup_{j \in J} S_j$ be a partition of S. Given a sequence $(a(s,n))_{s \in S}$ of families of noncommutative random variables $a(s,n) \in \mathcal{A}_n$, it is said that $(\{a(s,n) : s \in S_j\})_{j \in J}$ is *asymptotically free* (as $n \to \infty$) if the following two conditions hold true:

(i) There exists a linear functional μ ($\mu(\mathbf{1}) = 1$) on the noncommutative polynomial algebra $\mathbb{C}\langle X_s | s \in S\rangle$ such that for every $s_1, \dots, s_k \in S$, $k \in \mathbb{N}$,

$$\lim_{n\to\infty} \varphi_n(a(s_1, n)a(s_2, n)\cdots a(s_k, n)) = \mu(X_{s_1}X_{s_2}\cdots X_{s_k}).$$

Namely, $(a(s,n))_{s\in S}$ has the limit distribution μ in the sense of joint moments.

(ii) The family $(\mathbb{C}\langle X_s | s \in S_j\rangle)_{j\in J}$ of subalgebras of $\mathbb{C}\langle X_s | s \in S\rangle$ is free in the noncommutative probability space $(\mathbb{C}\langle X_s | s \in S\rangle, \mu)$.

Before stating Voiculescu's asymptotic freeness results for random matrices, let us introduce terminology for some standard random matrices.

1. A selfadjoint $n \times n$ random matrix $H(n)$ is said to be a *standard selfadjoint Gaussian* matrix (or *GUE* [61]) if $\{\Re H_{ij}(n) : 1 \le i \le j \le n\} \cup \{\Im H_{ij}(n) : 1 \le i < j \le n\}$ is an independent family of Gaussian random variables such that $E(H_{ij}(n)) = 0$ $(1 \le i \le j \le n)$, $E((\Re H_{ii}(n))^2) = 1/n$ $(1 \le i \le n)$ and $E((\Re H_{ij}(n))^2) = E((\Im H_{ij}(n))^2) = 1/(2n)$ $(1 \le i < j \le n)$.
2. An $n \times n$ random matrix $X(n)$ is said to be a *standard non-selfadjoint Gaussian* random matrix if $\{\Re X_{ij}(n), \Im X_{ij}(n) : 1 \le i, j \le n\}$ is an independent family of Gaussian random variables such that $E(X_{ij}(n)) = 0$ and $E((\Re X_{ij}(n))^2) = E((\Im X_{ij}(n))^2) = 1/(2n)$ $(1 \le i, j \le n)$.
3. An $n\times n$ unitary random matrix $U(n)$ is said to be *standard* if its distribution on the unitary group $\mathcal{U}(n)$ is the Haar probability measure.

THEOREM ([100, 108]). *Let S and T be sets and let $D(t,n) \in M_n$ $(t \in T)$ be constant matrices such that $\sup_n \|D(t,n)\| < +\infty$ $(t \in T)$ and $(D(t,n))_{t\in T}$ has the limit distribution.*

(1) *If $(H(s,n))_{s\in S}$ is an independent family of $n\times n$ standard selfadjoint Gaussian matrices, then*

$$\Big(\big(\{H(s,n)\}\big)_{s\in S}, \{D(t,n) : t \in T\}\Big)$$

is asymptotically free. Moreover, $\mu_{H(s,n)}$ converges to w_2.

(2) *If $(X(s,n))_{s\in S}$ is an independent family of $n \times n$ standard non-selfadjoint Gaussian matrices, then*

$$\Big(\big(\{X(s,n), X(s,n)^*\}\big)_{s\in S}, \{D(t,n) : t \in T\}\Big)$$

is asymptotically free. Moreover, $\mu_{X(s,n)}$ converges to the circular distribution (i.e., the uniform distribution on the unit disk).

(3) *If $(U(s,n))_{s\in S}$ is an independent family of $n \times n$ standard unitary random matrices, then*

$$\Big(\big(\{U(s,n), U(s,n)^*\}\big)_{s\in S}, \{D(t,n) : t \in T\}\Big)$$

is asymptotically free. Moreover, $\mu_{U(s,n)}$ converges to the uniform distribution on $\mathbb{T}$.

The above theorem says that the standard random matrices $H(n)$, $X(n)$, and $U(n)$ introduced above are asymptotic models of a semicircular element, a circular element, and a Haar unitary, respectively. Also, a (compound) Wishart matrix is a model of a (compound) free Poisson element ([53]). The asymptotic freeness was shown in [26] for random matrices more general than Gaussians and in [62] for random unitary matrices in subgroups of $\mathcal{U}(n)$. In [100, 26] constant matrices

$D(t,n)$ were assumed to be (block) diagonal; this restriction was removed in [108]. The inclusion of constant matrices is useful in applications to operator algebras, in particular, in the construction of interpolated free group factors mentioned in the next section. Another proof of the above theorem is found in [82]. The empirical spectral density is known to converge almost surely for various random matrices including the above-mentioned standard ones (see the recent [46], as well as many papers from the probability theory side). In accordance with this, it was proved in [51, 53] that the asymptotic freeness for random matrices holds in the stronger sense of almost sure convergence as well. The almost sure convergence was also shown in [90] for mixed moments of Gaussian random matrices.

Finally, we discuss the *large deviation principle* related to random matrices. The large deviation principle is a prominent theme of probability theory; see [25] for the general theory. Let $\mathcal{M}(\mathbb{R})$ denote the space of probability measures on $\mathbb{R}$ endowed with the weak topology (making $\mathcal{M}(\mathbb{R})$ a separable complete metric space). The empirical distribution on $\mathcal{M}(\mathbb{R})$ of an $n \times n$ standard selfadjoint Gaussian matrix $H(n)$ is defined by $P_n(\Gamma) := \mathbf{P}(R_{H(n)} \in \Gamma)$ for Borel subsets $\Gamma \subset \mathcal{M}(\mathbb{R})$. The *free entropy* of $\mu \in \mathcal{M}(\mathbb{R})$ was defined in [101] by the double integral

$$\Sigma(\mu) := \iint \log|x-y|\, d\mu(x)\, d\mu(y)$$

(this is nothing but the negative of the logarithmic energy familiar in potential theory). The joint density function on $\mathbb{R}^n$ of the eigenvalues of $H(n)$ is given as

$$\frac{(2\pi)^{-n/2} n^{n^2/2}}{\prod_{j=1}^n j!} \exp\left(-\frac{n}{2}\sum_{i=1}^n x_i^2\right) \prod_{i<j}(x_i - x_j)^2$$

(see [61]). From this, Ben Arous and Guionnet [6] proved the following large deviation result.

THEOREM ([6]). *The sequence (P_n) given above satisfies the large deviation principle in the scale n^{-2} with the good rate function*

$$I(\mu) := -\Sigma(\mu) + \frac{1}{2}\int x^2\, d\mu(x) - \frac{3}{4} \qquad (\mu \in \mathcal{M}(\mathbb{R})).$$

Moreover, w_2 is a unique minimizer of I with $I(w_2) = 0$.

This theorem asserts that $R_{H(n)}$ converges to w_2 exponentially fast. Similar large deviation results were proved also for unitary random matrices ([49]), non-selfadjoint Gaussian matrices ([7, 71]) and Wishart matrices ([50]). The logarithmic energy (or minus the free entropy) appears as the main term of rate functions from the form of joint eigenvalue density of these random matrices, and the minimizer of the rate function always coincides with the limit distribution of the random matrix (see [48]). Furthermore, using the method from the large deviation principle, one can prove ([53]) that if $\mu \in \mathcal{M}(\mathbb{R})$ is supported in $[-R, R]$, then for every $r \in \mathbb{N}$, $\varepsilon > 0$, the limit

$$\chi_R(\mu; r, \varepsilon)$$
$$:= \lim_{n\to\infty}\left[\frac{1}{n^2}\log\Lambda_n\big(\{A \in M_n^{sa} : |\mathrm{tr}_n(A^k) - m_k(\mu)| \le \varepsilon,\ k \le r\}\big) + \frac{1}{2}\log n\right]$$

exists, and

$$\inf_{r\in\mathbb{N},\,\varepsilon>0}\chi_R(\mu;r,\varepsilon)=\Sigma(\mu)+\frac{1}{2}\log(2\pi)+\frac{3}{4}.$$

Here Λ_n is the Lebesgue measure on M_n^{sa} ($\cong\mathbb{R}^{n^2}$) and $m_k(\mu)$ is the kth moment of μ. (The last formula was proved in [102], but the large deviation principle was not mentioned there.)

4. Free group factors and interpolated free group factors

Let a free family of selfadjoint noncommutative random variables $a_1,\dots,a_N$ be given in a W^*-probability space. If the distribution of each a_i is non-atomic, then a_i generates the von Neumann algebra $\mathcal{L}(\mathbb{Z})$, and so $\{a_1,\dots,a_n\}''$ is isomorphic to $\ast_{i=1}^{n}\mathcal{L}(\mathbb{Z})=\mathcal{L}(\mathbf{F}_n)$. This fact tells us that the three concepts of freeness, free product, and free group factors have the same essence. This is the reason why free probability theory is so useful in the analysis of free group factors. One of the biggest open problems on operator algebras since von Neumann would be the isomorphism problem of free group factors: "$\mathcal{L}(\mathbf{F}_n)\not\cong\mathcal{L}(\mathbf{F}_m)$ for $n\ne m$?" (Concerning free group C^*-algebras, their K-theory proved ([72]) that $C_r^*(\mathbf{F}_n)$ is not isomorphic to $C_r^*(\mathbf{F}_m)$ for $n\ne m$.) As mentioned below, the interpolated free group factors constructed independently by R. Rădulescu [79] and K. Dykema [28] produced much progress on this problem; however, the final solution is not yet achieved.

Let $\mathcal{M}$ be a type II_1 factor with tracial state τ, and let $B(\mathcal{H})$ be the type I_∞ factor with the usual trace Tr. For each $t\in(0,\infty)$ choose a projection $p\in\mathcal{M}\otimes B(\mathcal{H})$ with $(\tau\otimes\mathrm{Tr})(p)=t$, and fix a type II_1 factor $\mathcal{M}_t:=p(\mathcal{M}\otimes B(\mathcal{H}))p$, which is independent (up to an isomorphism) of the choice of the projection p. Since $\mathcal{M}_{t_1t_2}\cong(\mathcal{M}_{t_1})_{t_2}$ $(t_1,t_2>0)$, $\{t\in(0,\infty):\mathcal{M}_t\cong\mathcal{M}\}$ becomes a subgroup of the multiplicative group $(0,\infty)$, and it is called the *fundamental group* of $\mathcal{M}$. For instance, the fundamental group of the hyperfinite II_1 factor is the whole $(0,\infty)$.

The *interpolated free group factors* are a one-parameter family of type II_1 factors $\mathcal{L}(\mathbf{F}_r)$ $(1<r\le\infty)$ that satisfy the following three properties (the notation $\mathbf{F}_r$ by itself does not make sense):

(i) When $r=n\in\{2,3,\dots,\infty\}$, $\mathcal{L}(\mathbf{F}_r)\cong\mathcal{L}(\mathbf{F}_n)$.
(ii) Compression formula: $\mathcal{L}(\mathbf{F}_r)_t\cong\mathcal{L}(\mathbf{F}_{1+(r-1)/t^2})$ for every $1<r\le\infty$ and $0<t<\infty$.
(iii) Addition formula: $\mathcal{L}(\mathbf{F}_r)*\mathcal{L}(\mathbf{F}_{r'})\cong\mathcal{L}(\mathbf{F}_{r+r'})$ for every $1<r,r'\le\infty$.

The properties (i) and (ii) imply that $\mathcal{L}(\mathbf{F}_r)\cong\mathcal{L}(\mathbf{F}_n)_t$ for $1<r<\infty$ and $t=((n-1)/(r-1))^{1/2}$, so the interpolated free group factors $\mathcal{L}(\mathbf{F}_r)$ $(1<r\le\infty)$ are uniquely determined, if they exist. The constructions in [79] and [28] are a bit different, but in both a semicircular system is used, and the Gaussian random matrices discussed in Section 3 play an essential role. Here let us mention the definition in [28]. In a tracial W^*-probability space $(\mathcal{M},\tau)$ let $(a_s)_{s\in S}$ be a semicircular system (S being an infinite set), and let R be a copy of the hyperfinite II_1 factor such that R is free from $(a_s)_{s\in S}$. For each $1<r\le\infty$, choose projections $p_s\in R$ $(s\in S)$ so that $r=1+\sum_{s\in S}\tau(p_s)^2$. Then it was proved in [28] that

$$\mathcal{L}(\mathbf{F}_r):=(R\cup\{p_sa_sp_s\}_{s\in S})''$$

is a type II_1 factor determined independently of the choice of $(p_s)_{s\in S}$. To prove the properties (i)–(iii), the following important formulas on free products were used.

THEOREM ([26, 28]). *For* $n = 1, 2, \ldots, \infty$, $R * \mathcal{L}(\mathbf{F}_n) \cong \mathcal{L}(\mathbf{F}_{n+1})$ *(with* $\mathbf{F}_1 = \mathbb{Z}$*). Moreover,* $R * R \cong R * \mathcal{L}(\mathbb{Z}) \cong \mathcal{L}(\mathbf{F}_2)$.

The following results can easily be checked from the properties of interpolated free group factors.

(1) The fundamental group of $\mathcal{L}(\mathbf{F}_\infty)$ is $(0, \infty)$.
(2) For every $1 < r, s < \infty$, $\mathcal{L}(\mathbf{F}_r) \otimes B(\mathcal{H}) \cong \mathcal{L}(\mathbf{F}_s) \otimes B(\mathcal{H})$ ($B(\mathcal{H})$ is of type I_∞).
(3) For every $1 < r, s < \infty$, the isomorphism class of $\mathcal{L}(\mathbf{F}_r)\otimes\mathcal{L}(\mathbf{F}_s)$ is determined only by $(r-1)(s-1)$. In particular, $\mathcal{L}(\mathbf{F}_r) \otimes \mathcal{L}(\mathbf{F}_\infty) \cong \mathcal{L}(\mathbf{F}_s) \otimes \mathcal{L}(\mathbf{F}_\infty)$.

Furthermore, concerning the isomorphism problem of $\mathcal{L}(\mathbf{F}_r)$, one of the following two extreme cases occurs:

1° $\mathcal{L}(\mathbf{F}_r)$ $(1 < r < \infty)$ are all isomorphic and the fundamental group of $\mathcal{L}(\mathbf{F}_r)$ is $(0, \infty)$ for every $1 < r < \infty$.
2° Any two of $\mathcal{L}(\mathbf{F}_r)$ $(1 < r < \infty)$ are mutually nonisomorphic and the fundamental group of $\mathcal{L}(\mathbf{F}_r)$ is $\{1\}$ for every $1 < r < \infty$.

Before the construction of interpolated free group factors, the above results (1) and (2) were proved in [76] and [78], respectively. It was proved earlier, in [99], that the fundamental group of $\mathcal{L}(\mathbf{F}_\infty)$ contains all positive rationals. Furthermore, a remarkable result in [77] is that there exists a one-parameter family $(\alpha_t)_{t>0}$ of trace-scaling automorphisms of $\mathcal{L}(\mathbf{F}_\infty)\otimes B(\mathcal{H})$ (i.e., $(\tau\otimes\mathrm{Tr})\circ\alpha_t = t(\tau\otimes\mathrm{Tr})$, $t > 0$). This includes (1).

Finally, which statement, 1° or 2°, holds true? The existence of the interpolation with continuous parameter of the free group factors $\mathcal{L}(\mathbf{F}_n)$ with discrete parameter may suggest that $\mathcal{L}(\mathbf{F}_n)$ are all isomorphic. On the other hand, Voiculescu introduced the *free entropy dimension* $\delta(a_1, \ldots, a_n)$ as a kind of derivative of the free entropy $\chi(a_1, \ldots, a_N)$ (discussed in Section 6), and if this $\delta(a_, \ldots, a_N)$ happens to be a lower semicontinuous function of $(a_1, \ldots, a_N)$ in the strong topology, then necessarily $\mathcal{L}(\mathbf{F}_n) \not\cong \mathcal{L}(\mathbf{F}_m)$ for $n \neq m$. In this way, there are two opposite signs, so that it seems difficult to conjecture either 1° or 2°.

5. Free products and amalgamated free products of operator algebras

Since we gave the definition of free products in Section 1, let us here start with the definition of amalgamated free products. Let $\mathcal{A}_1, \mathcal{A}_2$ be two algebras having a common subalgebra $\mathcal{B}$ (containing $\mathbf{1}$), and let $\Phi_i : \mathcal{A}_i \to \mathcal{B}$ $(i = 1, 2)$ be linear maps satisfying $\Phi_i(\mathbf{1}) = \mathbf{1}$ and the bimodule property $\Phi_i(b_1 a b_2) = b_1 \Phi_i(a) b_2$ $(a \in \mathcal{A}_i$, $b_1, b_2 \in \mathcal{B})$. Letting $\mathcal{A}_i^0 := \ker\Phi_i = \{a - \Phi_i(a) : a \in \mathcal{A}_i\}$, define

$$\mathcal{A} := \mathcal{B} \oplus \bigoplus_{n=1}^{\infty}\left(\bigoplus_{i_1\neq i_2\neq\cdots\neq i_n} \mathcal{A}_{i_1}^0 \otimes_{\mathcal{B}} \mathcal{A}_{i_2}^0 \otimes_{\mathcal{B}} \cdots \otimes_{\mathcal{B}} \mathcal{A}_{i_n}^0\right),$$

where $\otimes_{\mathcal{B}}$ denotes the relative tensor product with respect to $\mathcal{B}$, so that $a_1 b\otimes_{\mathcal{B}} a_2 = a_1\otimes_{\mathcal{B}} b a_2$ $(b \in \mathcal{B})$. Then $\mathcal{A}$ becomes an algebra, similarly to the case of free products. Also, define $\Phi : \mathcal{A} \to \mathcal{B}$ by $\Phi(b + \sum a_1 a_2 \cdots a_n) := b$. This $(\mathcal{A}, \Phi)$ is called the *amalgamated free product* of $(\mathcal{A}_1, \Phi_1)$ and $(\mathcal{A}_2, \Phi_2)$, and we write $\mathcal{A} = \mathcal{A}_1 *_{\mathcal{B}} \mathcal{A}_2$.

The particular case of $\mathcal{B} = \mathbb{C}\mathbf{1}$ reduces to the case of free products. When $\mathcal{A}_i$ ($\supset \mathcal{B}$) are C^*-algebras and $\Phi_i : \mathcal{A} \to \mathcal{B}$ are conditional expectations (i.e., norm 1 projections), the *reduced* C^**-amalgamated free product* of $(\mathcal{A}_i, \Phi_i)$ is the C^*-algebra generated by the representations of $\mathcal{A}_i$'s on the free product of the Hilbert $\mathcal{B}$-modules made by $(\mathcal{A}_i, \Phi_i)$'s ([96]). Moreover, the amalgamated free product of von Neumann algebras was first introduced in [74] in the case with respect to normal tracial states. The precise definition of the amalgamated free product of general von Neumann algebras is found in [92]. As free probability theory is related to free products, a noncommutative probability related to amalgamated free products can be developed; it extends free probability theory to the operator-valued ($\mathcal{B}$-valued) theory ([87, 103]).

First, let us mention recent results on free products of von Neumann algebras. By examining the free products of finite-dimensional algebras, the hyperfinite II_1 factor, interpolated free group factors, etc., Dykema [27] introduced the notion of the *free dimension* $\mathrm{fdim}(\mathcal{M})$ for these von Neumann algebras so that the addition formula $\mathrm{fdim}(\mathcal{M}_1 * \mathcal{M}_2) = \mathrm{fdim}(\mathcal{M}_1) + \mathrm{fdim}(\mathcal{M}_2)$ is satisfied. (But the free dimension is not well defined if the interpolated free group factors happen to be all isomorphic.) In the first papers [5, 29] dealing with the free product of von Neumann algebras with respect to non-tracial states, it was proved that if φ_1 or φ_2 is not tracial, then $(\mathcal{M}, \varphi) := (\mathcal{M}_1, \varphi_1) * (\mathcal{M}_2, \varphi_2)$ is in almost all cases a type III factor (even when $\mathcal{M}_1, \mathcal{M}_2$ are finite-dimensional algebras). Starting from the equality $\sigma_t^\varphi = \sigma_t^{\varphi_1} * \sigma_t^{\varphi_2}$ for modular automorphism groups, the formula $T(\mathcal{M}) = \{t \in \mathbb{R} : \sigma_t^{\varphi_1} = \mathrm{id},\ \sigma_t^{\varphi_2} = \mathrm{id}\}$ for Connes' T-set was obtained under a rather mild assumption. A condition for $\mathcal{M}$ to be *full* (equivalent to $\mathcal{M}$ not having property Γ when $\mathcal{M}$ is a type II_1 factor) was given in [5]. Furthermore, in [31] the structure of $(\mathcal{M}_1, \varphi_1) * (\mathcal{M}_2, \varphi_2)$ was completely determined in some cases, including the case where $\mathcal{M}_1$ and $\mathcal{M}_2$ are finite-dimensional. The free product von Neumann algebras treated there are type III_λ factors $(0 < \lambda \le 1)$ except for possible finite-dimensional summands, and their centralizer with respect to the free product state is isomorphic to $\mathcal{L}(\mathbf{F}_\infty)$. Related results are also in [93]. On the other hand, [39] extended results of [73] via free product techniques.

The method of (amalgamated) free products is sometimes useful for constructing (sub)factors or their actions having a prescribed property. However, it may be noted that the factors thus constructed are non-hyperfinite ones akin to free group factors; moreover, it seems that they do not have property Γ or are full. The first major success in this direction was made by Popa [74], where an irreducible type II_1 subfactor $N^s \subset M^s$ of index s was constructed for each $s \in \{4\cos^2\frac{\pi}{n} : n \ge 3\} \cup [4, \infty)$ (see also [16, 75] for related work). In fact, $N^s \subset M^s$ was made in $(Q \otimes R_\lambda) *_{R_\lambda} R$, where $R_\lambda \subset R$ is the Jones subfactor of index $s = \lambda^{-1}$ and Q is a nonatomic von Neumann algebra ($L^\infty[0,1]$ is enough for Q). Furthermore, in [79], in the case where

$$\begin{matrix} B & \subset & A \\ \cup & & \cup \\ D & \subset & C \end{matrix}$$

is a commuting square of finite-dimensional algebras and A has a λ-Markov trace with respect to $C \subset A$, a type II_1 subfactor of index λ^{-1} was constructed as follows:

$$(\mathcal{N} \subset \mathcal{M}) := ((\mathcal{L}(\mathbf{F}_\infty) \otimes D) *_D C \subset (\mathcal{L}(\mathbf{F}_\infty) \otimes B) *_B A).$$

Concerning this $\mathcal{N} \subset \mathcal{M}$, the use of random matrix models proved that $\mathcal{N} \cong \mathcal{M} \cong \mathcal{L}(\mathbf{F}_\infty)$. Subfactors of $\mathcal{L}(\mathbf{F}_\infty)$ were also constructed in [30] by using the method of amalgamated free products. On the other hand, it is noteworthy that the free product was used in [91] to show that the compact quantum group $SU_q(n)$ enjoys a minimal coaction on a factor (more precisely, a full type III_{q^2} factor). Amalgamated free products of von Neumann algebras over a common Cartan algebra were recently investigated in [92] and [56]; in fact, the notion of free products for measurable relations was introduced in [56].

As for (amalgamated) free products of C^*-algebras, it was proved in [34] that if $(\mathcal{A}_i, \varphi_i)$ are C^*-algebras with faithful tracial states φ_i and they satisfy the Avitzour condition, then the reduced free product C^*-algebra $(\mathcal{A}_1, \varphi_1) * (\mathcal{A}_2, \varphi_2)$ has stable rank (a notion of dimension for C^*-algebras) 1. In particular, if G_1, G_2 are discrete groups such that $|G_1| \ge 2$ and $|G_2| \ge 3$, then $C_r^*(G_1 * G_2)$ (for instance, $C_r^*(\mathbf{F}_n)$, $n \ge 2$) has stable rank 1 (see also [60] in this connection). Among many papers of Dykema are [32] and [33]; it was proved in the latter (also [37]) that the reduced amalgamated free product C^*-algebra of exact C^*-algebras is exact. Moreover, free shift dynamical systems and their dynamical entropy have been studied in [20, 23, 24, 89].

To conclude this section we summarize Shlyakhtenko's results in [83, 84]. Let U_t be a one-parameter group of orthogonal transformations on a real Hilbert space $\mathcal{H}_\mathbb{R}$, whose linear extension to $\mathcal{H}_\mathbb{C} := \mathcal{H}_\mathbb{R} \oplus \mathrm{i}\,\mathcal{H}_\mathbb{R}$ is written as $U_t = A^{\mathrm{i}t}$ ($A \ge 0$ is unbounded in general). On the full Fock space $\mathcal{F}(\mathcal{H}_\mathbb{C})$, define $\Gamma(\mathcal{H}_\mathbb{R}, U_t) := C^*\big(\{s(\sqrt{2}(\mathbf{1} + A^{-1})^{-1/2}h) : h \in \mathcal{H}_\mathbb{R}\}\big)$ ($s(h)$ was defined in Section 2), $\alpha_t := \mathrm{Ad}\,\mathcal{F}(U_t) \in \mathrm{Aut}\,\Gamma(\mathcal{H}_\mathbb{R}, U_t)$ and $\varphi_U := \langle \cdot\,\Phi, \Phi\rangle$. Then φ_U is an α_t-KMS state; $\Gamma(\mathcal{H}_\mathbb{R}, U_t)$ and φ_U are regarded as free probabilistic analogues of the CAR algebra and its quasi-free state; so $\Gamma(\mathcal{H}_\mathbb{R}, U_t)''$ can be regarded as a free analogue of an Araki-Wood factor. If $U_t \equiv \mathbf{1}$, then $\Gamma(\mathcal{H}_\mathbb{R}, U_t)'' \cong \mathcal{L}(\mathbf{F}_{\dim \mathcal{H}_\mathbb{R}})$ (an example of Section 2). The crucial case arises if

$$U_t = \begin{pmatrix} \cos(t\log\lambda) & -\sin(t\log\lambda) \\ \sin(t\log\lambda) & \cos(t\log\lambda) \end{pmatrix} \quad (\text{period } 2\pi/\log\lambda)$$

on the two-dimensional $\mathcal{H}_\mathbb{R} = \mathbb{R}^2$. In this case, concerning $(T_\lambda, \varphi_\lambda) := (\Gamma(\mathcal{H}_\mathbb{R}, U_t)'', \varphi_U)$ for $0 < \lambda < 1$, we have the following.

THEOREM ([83]). $(T_\lambda, \varphi_\lambda) * (\mathcal{L}(\mathbb{Z}), \tau) \cong (T_\lambda, \varphi_\lambda) \cong (M_2, \psi_\lambda) * (\mathcal{L}(\mathbb{Z}), \tau)$, *where*

$$\psi_\lambda := \mathrm{Tr}\left(\begin{bmatrix} \frac{\lambda}{1+\lambda} & 0 \\ 0 & \frac{1}{1+\lambda} \end{bmatrix}\right).$$

From this and [80] it is seen that $(T_\lambda, \varphi_\lambda)$ is a type III_λ factor whose core is $\mathcal{L}(\mathbf{F}_\infty) \otimes B(\mathcal{H})$.

THEOREM ([83, 84]). *If U_t is periodic with period $2\pi/\log\lambda$ ($0 < \lambda < 1$), then $\Gamma(\mathcal{H}_\mathbb{R}, U_t)''$ is a type III_λ factor. If U_t is not periodic, then $\Gamma(\mathcal{H}_\mathbb{R}, U_t)''$ is a type III_1 factor. Moreover, when U_t ($= A^{\mathrm{i}t}$) $\not\equiv \mathbf{1}$ is almost periodic, the isomorphism class of the arising factor is determined by the subgroup of the multiplicative group $(0, \infty)$ generated by the point spectrum of A.*

When $(\mathcal{H}_\mathbb{R}, U_t) := (L^2(\mathbb{R}; \mathbb{R}), f \mapsto f(\cdot + t))$, the core (i.e., the Takesaki dual) of the type III_1 factor $\Gamma(\mathcal{H}_\mathbb{R}, U_t)''$ is $\mathcal{L}(\mathbf{F}_\infty) \otimes B(\mathcal{H})$ and the dual action on the core coincides with the trace-scaling action in [77]. Furthermore, for a type II_1 factor $\mathcal{M}$, a compression formula for $\mathcal{M} * \mathcal{L}(\mathbf{F}_r)$ was given in [84], implying that the fundamental group of $\mathcal{M}$ is included in that of $\mathcal{M} * \mathcal{L}(\mathbf{F}_\infty)$. This result extends (1) of Section 4 because $R * \mathcal{L}(\mathbf{F}_\infty) \cong \mathcal{L}(\mathbf{F}_\infty)$. (A more general compression formula for free products of type II_1 factors was recently obtained in [36].)

6. Free entropy and applications to operator algebras

In this section we work in a tracial W^*-probability space $(\mathcal{M}, \tau)$. The set of selfadjoint elements in $\mathcal{M}$ is denoted by $\mathcal{M}^{sa}$. Looking at the asymptotic behavior of the Boltzmann-Gibbs entropy of random matrices, Voiculescu proposed first in [101] that the double logarithmic integral $\Sigma(\mu)$ given in Section 3 should be the free entropy of a probability distribution μ. Later, in [102] he found the formula given at the end of Section 3 and introduced his multi-variable free entropy as follows.

Let $(a_1, \dots, a_N)$ be an N-tuple of elements in $\mathcal{M}^{sa}$. For a multi-index $k = (k_1, \dots, k_s)$, $1 \le k_i \le N$, write $|k| := s$ and $m_k(a_1, \dots, a_N) := \tau(a_{k_1} \cdots a_{k_s})$ for the corresponding joint moment. Similarly, write $m_k(A_1, \dots, A_N) := \mathrm{tr}_n(A_{k_1} \cdots A_{k_s})$ for $A_1, \dots, A_N \in M_n$. For $n, r \in \mathbb{N}$, $\varepsilon > 0$ and $R > 0$, define the following set of N-tuples of $n \times n$ selfadjoint matrices approximating $(a_1, \dots, a_n)$ in joint moments:

$$\begin{aligned}\Gamma_R(a_1, \dots, a_N; n, r, \varepsilon) := \big\{ &(A_1, \dots, A_N) \in (M_n^{sa})^N : \|A_i\| \le R, \\ &|m_k(A_1, \dots, A_N) - m_k(a_1, \dots, a_N)| \le \varepsilon,\ |k| \le r \big\}.\end{aligned}$$

Then, the *free entropy* of $(a_1, \dots, a_N)$ is

$$\begin{aligned}&\chi(a_1, \dots, a_N) \\ &:= \sup_{R>0} \inf_{r \in \mathbb{N},\, \varepsilon > 0} \limsup_{n \to \infty} \left[\frac{1}{n^2} \log \Lambda_n^{\otimes N}(\Gamma_R(a_1, \dots, a_N; n, r, \varepsilon)) + \frac{N}{2} \log n \right].\end{aligned}$$

This $\chi(a_1, \dots, a_N)$ is sometimes called the *microstates free entropy*, as matrices approximating $(a_1, \dots, a_N)$ are viewed as microstates. In particular, in the single variable case, $\chi(a)$ coincides with $\Sigma(\mu)$ in Section 3 up to a constant, where μ is the distribution of a.

The following are fundamental properties of $\chi(a_1, \dots, a_N)$ presented in [102, 106].

1. Subadditivity:
$$\begin{aligned}\chi(a_1, \dots, a_N) &\le \chi(a_1, \dots, a_L) + \chi(a_{L+1}, \dots, a_N), \\ \chi(a_1, \dots, a_N) &\le \chi(a_1) + \dots + \chi(a_N).\end{aligned}$$
2. Upper semicontinuity: If $a_{m,i} \in \mathcal{M}^{sa}$ $(1 \le i \le N,\ m \in \mathbb{N})$, $\sup_m \|a_{m,i}\| < +\infty$ and $(a_{m,1}, \dots, a_{m,N})$ converges to $(a_1, \dots, a_N)$ in joint moments, then $\chi(a_1, \dots, a_N) \ge \limsup_{m \to \infty} \chi(a_{m,1}, \dots, a_{m,N})$.
3. If $b_1 = a_1$, $b_i - a_i \in \{a_1, \dots, a_{i-1}\}''$ $(1 \le i \le N)$, then $\chi(a_1, \dots, a_N) = \chi(b_1, \dots, b_N)$.
4. Change of variable formula in power series: If $F_i(X_1, \dots, X_N)$ and $G_i(X_1, \dots, X_N)$ $(1 \le i \le N)$ are noncommutative selfadjoint power series and they satisfy $G_i(F_1(X_1, \dots, X_N), \dots, F_N(X_1, \dots, X_N)) = X_i$ $(1 \le i \le N)$, then
$$\begin{aligned}&\chi(F_1(a_1, \dots, a_N), \dots, F_N(a_1, \dots, a_N)) \\ &\qquad = \chi(a_1, \dots, a_N) + \log |\mathcal{J}|(F)(a_1, \dots, a_N)\end{aligned}$$
under a certain assumption on the radii of convergence of F_i and G_i, where $|\mathcal{J}|(F)(a_1, \dots, a_N)$ is the Jacobian of $F := (F_1, \dots, F_N)$ at $(a_1, \dots, a_N)$ defined via the Fuglede-Kadison determinant.

5. Separate change of variable formula: If $\chi(a_i) > -\infty$ and f_i $(1 \le i \le N)$ are strictly increasing continuous functions on $\mathbb{R}$, then

$$\chi(f_1(a_1), \dots, f_N(a_N)) = \chi(a_1, \dots, a_N) + \sum_{i=1}^{N} \left[\chi(f_i(a_i)) - \chi(a_i)\right].$$

6. Maximization: If $\tau(a_1^2 + \cdots + a_N^2) \le C$, then $\chi(a_1, \dots, a_N) \le \frac{N}{2} \log(2\pi e C/N)$, and the equality holds if and only if $a_1, \dots, a_N$ are free and all a_i have the same distribution $w_{2\sqrt{C/N}}$.
7. Additivity: If $a_1, \dots, a_N$ are free, then $\chi(a_1, \dots, a_N) = \chi(a_1) + \cdots + \chi(a_N)$. Conversely, if this equality holds with $\chi(a_i) > -\infty$ $(1 \le i \le N)$, then $a_1, \dots, a_N$ are free.

The free entropy $\hat{\chi}(a_1, \dots, a_N)$ for non-selfadjoint $a_1, \dots, a_N \in \mathcal{M}$ can be defined by slightly modifying the above definition; in fact, it is easy to show that $\hat{\chi}(a_1, \dots, a_N) = \chi(b_1, c_1, \dots, b_N, c_N)$ when $a_i = b_i + ic_i$ $(b_i, c_i \in \mathcal{M}^{sa})$. The free entropy $\chi_u(u_1, \dots, u_N)$ for unitaries $u_1, \dots, u_N \in \mathcal{M}$ is also introduced. For a multi-index with signs $k = (\pm k_1, \dots, \pm k_s)$, $1 \le k_i \le N$, writing $m_k(u_1, \dots, u_N) := \tau(u_{k_1}^{\varepsilon_1} \cdots u_{k_s}^{\varepsilon_s})$ (ε_i being the sign of k_i) and similarly $m_k(U_1, \dots, U_N)$ for unitary matrices $U_1, \dots, U_N$, we define

$$\Gamma_u(u_1, \dots, u_N; n, r, \varepsilon) := \big\{(U_1, \dots, U_N) \in \mathcal{U}(n)^N : \\ |m_k(U_1, \dots, U_N) - m_k(u_1, \dots, u_N)| \le \varepsilon,\ |k| \le r\big\},$$

$$\chi_u(u_1, \dots, u_N) := \inf_{r \in \mathbb{N},\, \varepsilon > 0} \limsup_{n \to \infty} \frac{1}{n^2} \log \gamma_n^{\otimes N}(\Gamma_u(u_1, \dots, u_N; n, r, \varepsilon)),$$

where γ_n is the Haar probability measure on the unitary group $\mathcal{U}(n)$. In the single variable case, if μ is the distribution on $\mathbb{T}$ of u, then we have

$$\chi_u(\mu) = \iint_{\mathbb{T}^2} \log|\zeta - \eta|\, du(\zeta)\, du(\eta).$$

The relation among the three kinds of free entropies χ, $\hat{\chi}$ and χ_u is elaborated upon in the next theorem.

THEOREM ([52]). *If $a_i = u_i h_i$ $(1 \le i \le N)$ are the polar decompositions with unitaries u_i and $h_i \ge 0$, then*

$$\hat{\chi}(a_1, \dots, a_N) \le \chi_u(u_1, \dots, u_N) + \chi(h_1^2, \dots, h_N^2) + \frac{N}{2}\left(\log\frac{\pi}{2} + \frac{3}{2}\right).$$

Furthermore, if $\{u_1, \dots, u_N, u_1^*, \dots, u_N^*\}$, $h_1, \dots, h_N$ are free, then equality holds in the above.

This formula is employed to prove that χ_u has the same additivity property as χ, and it is also applied to solve some maximization problems for $\hat{\chi}$. For instance, if the distribution of a^*a is fixed, then $\hat{\chi}(a)$ is maximal if and only if a is R-diagonal ([52, 66], also [65]).

Remarkably enough, many long-standing open problems on free group factors have been solved by making use of the free entropy or the free entropy dimension. Let us briefly summarize them.

1. The interpolated free group factors $\mathcal{L}(\mathbf{F}_r)$ $(1 < r \le \infty)$ have no *Cartan subalgebras* ([105]). This means that $\mathcal{L}(\mathbf{F}_r)$ cannot be constructed by the measure space construction via a measurable relation ([38]). In fact, it

was more strongly proved in [105] that $\mathcal{L}(\mathbf{F}_r)$ has no normal nonatomic hyperfinite subalgebras. It was also shown that if $\chi(a_1, \dots, a_N) > -\infty$ $(N \geq 2)$, then $\{a_1, \dots, a_N\}''$ is a type II_1 factor without property Γ.

2. For each $2 \leq r < \infty$, $\mathcal{L}(\mathbf{F}_r)$ has no *simple* maximal abelian subalgebras ([40]).
3. For each $n = 2, 3, \dots$, $\mathcal{L}(\mathbf{F}_n)$ is *prime*, i.e., it is not the tensor product of two type II_1 factors ([41]).

In [42], it was proved by the free entropy technique that $\mathcal{L}(\mathbf{F}_n)$ is not thin if $n \geq 3$. (A type II_1 factor $\mathcal{M}$ is *thin* if $\mathcal{M} = \overline{\mathrm{span}}(R_1 R_2)$ for some hyperfinite subfactors R_1, R_2.) Furthermore, the above result in [41] was extended in [88] to show that any subfactor of $\mathcal{L}(\mathbf{F}_r)$ $(1 < r \leq \infty)$ of finite index (in particular, $\mathcal{L}(\mathbf{F}_r)$ itself) is prime. The recent papers [110, 43] are concerned with the free entropy dimension for sets of generators of some property T factors, showing that property T factors behave differently from free group factors.

In [107], the (*relative*) *free information* $\Phi^*(a_1, \dots, a_N : B)$ with respect to a subalgebra B, a free analogue of the Fisher information, was introduced by using the notion of *noncommutative Hilbert transform*, and the (*relative*) *free entropy* $\chi^*(a_1, \dots, a_N : B)$ with respect to B in the microstates-free approach was defined as a certain integral of the above free information. In the single variable case, $\chi(a) = \chi^*(a : \mathbb{C}\mathbf{1})$ holds, but it is not known whether $\chi(a_1, \dots, a_N) = \chi^*(a_1, \dots, a_N : \mathbb{C}\mathbf{1})$ holds in general or not. The extensions of Φ^* and χ^* to the setting of operator-valued free probability were treated in [85].

Concluding remarks

We have reviewed recent developments on free probability theory and free products of operator algebras. Our expositions are not complete in many points because of limitations of space, but the reader may consult the monographs [111] and [53]. (Algebraic) quantum probability theory, including free probability, is discussed in a very recent text [1]. Needless to say, free probability theory is one of the central subjects of current operator algebras. But we hope that people from other branches would recognize it as a proper theory in the probability world. Finally, we are grateful to Yoshimichi Ueda for his instruction about the literature on free products of operator algebras, and to the referee for his useful comments.

References

[1] L. Accardi and H. Obata, *Introduction to Algebraic Probability Theory–Various Concepts of Independence*, Graduate School Lectures in Math. **2**, Nagoya Univ., 1999 (in Japanese).

[2] M. Akiyama and H. Yoshida, *The distributions for linear combinations of a free family of projections and their orthogonal polynomials*, preprint.

[3] L. Arnold, *On the asymptotic distribution of the eigenvalues of random matrices*, J. Math. Anal. Appl. **20** (1967), 262–268.

[4] D. Avitzour, *Free products of C^*-algebras*, Trans. Amer. Math. Soc. **271** (1982), 423–435.

[5] L. Barnett, *Free product von Neumann algebras of type III*, Proc. Amer. Math. Soc. **123** (1995), 543–553.

[6] G. Ben Arous and A. Guionnet, *Large deviation for Wigner's law and Voiculescu's non-commutative entropy*, Probab. Theory Related Fields **108** (1997), 517–542.

[7] G. Ben Arous and O. Zeitouni, *Large deviations from the circular law*, ESAIM: Probability and Statistics **2** (1998), 123–134.

[8] H. Bercovici and D. Voiculescu, *Lévy-Hinčin type theorems for multiplicative and additive free convolution*, Pacific J. Math. **153** (1992), 217–248.

[9] H. Bercovici and D. Voiculescu, *Free convolution of measures with unbounded support*, Indiana Univ. Math. J. **42** (1993), 733–773.
[10] H. Bercovici and D. Voiculescu, *Superconvergence to the central limit and failure of the Cramér theorem for free random variables*, Probab. Theory Related Fields **102** (1995), 215–222.
[11] P. Biane, *Free Brownian motion, free stochastic calculus and random matrices*, in *Free Probability Theory*, D. V. Voiculescu (ed.), Fields Inst. Commun. **12**, Amer. Math. Soc., Providence, RI, 1997, pp. 1–19.
[12] P. Biane, *Processes with free increments*, Math. Z. **227** (1998), 143–174.
[13] P. Biane, *Representations of symmetric groups and free probability*, Adv. Math. **138** (1998), 126–181.
[14] P. Biane and R. Speicher, *Stochastic calculus with respect to free Brownian motion and analysis on Wigner space*, Probab. Theory Related Fields **112** (1998), 373–409.
[15] P. Biane and R. Speicher, *Free diffusions, free entropy and free Fisher information*, preprint.
[16] F. Boca, *On the method of constructing irreducible finite index subfactors of Popa*, Pacific J. Math. **161** (1993), 201–231.
[17] M. Bozejko, M. Leinert, and R. Speicher, *Convolution and limit theorems for conditionally free random variables*, Pacific J. Math. **175** (1996), 357–388.
[18] M. Bozejko and R. Speicher, *An example of a generalized Brownian motion I*, Comm. Math. Phys. **137** (1991), 519–531.
[19] M. Bozejko and R. Speicher, *Interpolation between bosonic and fermionic relations given by generalized Brownian motions*, Math. Z. **222** (1996), 135–160.
[20] N. Brown and M. Choda, *Approximation entropies in crossed products with an application to free shifts*, preprint.
[21] D. I. Cartwright and P. M. Soardi, *Random walks on free products, quotients and amalgams*, Nagoya Math. J. **102** (1986), 163–180.
[22] W.-M. Ching, *Free products of von Neumann algebras*, Trans. Amer. Math. Soc. **178** (1973), 147–163.
[23] M. Choda, *Reduced free products of completely positive maps and entropy for free products of automorphisms*, Publ. Res. Inst. Math. Sci. **32** (1996), 371–382.
[24] M. Choda and T. Natsume, *Reduced C^*-crossed products by free shifts*, Ergodic Theory Dynam. Systems **18** (1998), 1075–1096.
[25] A. Dembo and O. Zeitouni, *Large Deviation Techniques and Applications*, 2nd Edition, Springer-Verlag, New York, 1998.
[26] K. Dykema, *On certain free product factors via an extended matrix model*, J. Funct. Anal. **112** (1993), 31–60.
[27] K. Dykema, *Free products of hyperfinite von Neumann algebras and free dimension*, Duke Math. J. **69** (1993), 97–119.
[28] K. Dykema, *Interpolated free group factors*, Pacific J. Math. **163** (1994), 123–135.
[29] K. J. Dykema, *Factoriality and Connes' invariant $T(\mathcal{M})$ for free products of von Neumann algebras*, J. Reine Angew. Math. **450** (1994), 159–180.
[30] K. J. Dykema, *Amalgamated free products of multi-matrix algebras and a construction of subfactors of a free group factor*, Amer. J. Math. **117** (1995), 1555–1602.
[31] K. J. Dykema, Free products of finite dimensional and other von Neumann algebras with respect to non-tracial states, in *Free Probability Theory*, D. V. Voiculescu (ed.), Fields Inst. Commun. **12**, Amer. Math. Soc., 1997, pp. 41–88.
[32] K. J. Dykema, *Simplicity and the stable rank of some free product C^*-algebras*, Trans. Amer. Math. Soc. **351** (1999), 1–40.
[33] K. J. Dykema, *Exactness of reduced amalgamated free product C^*-algebras*, preprint.
[34] K. Dykema, U. Haagerup, and M. Rørdam, *The stable rank of some free product C^*-algebras*, Duke Math. J. **90** (1997), 95–121; Correction, *ibid.* **94** (1998), 213.
[35] K. Dykema and A. Nica, *On the Fock representation of the q-commutation relations*, J. Reine Angew. Math. **440** (1993), 201–212.
[36] K. Dykema and F. Rădulescu, *Compressions of free products of von Neumann algebras*, Math. Ann. **316** (2000), 61–82.
[37] K. J. Dykema and D. Shlyakhtenko, *Exactness of Cuntz-Pimsner C^*-algebras*, preprint.
[38] J. Feldman and C. C. Moore, *Ergodic equivalence relations, cohomology, and von Neumann algebras*. I, II, Trans. Amer. Math. Soc. **234** (1977), 289–324, 325–359.

[39] L. Ge, *On maximal injective subalgebras of factors*, Adv. Math. **118** (1996), 34–70.

[40] L. Ge, *Applications of free entropy to finite von Neumann algebras*, Amer. J. Math. **119** (1997), 467–485.

[41] L. Ge, *Applications of free entropy to finite von Neumann algebras, II*, Ann. of Math. (2) **147** (1998), 143–157.

[42] L. Ge and S. Popa, *On some decomposition properties for factors of type II_1*, Duke Math. J. **94** (1998), 79–101.

[43] L. Ge and J. Shen, *Applications of free entropy to finite von Neumann algebras, III*, preprint.

[44] U. Haagerup, *Quasitraces on exact C^*-algebras are traces*, handwritten manuscript.

[45] U. Haagerup, *On Voiculescu's R- and S-transforms for free non-commutative random variables*, in *Free Probability Theory*, D. V. Voiculescu (ed.), Fields Inst. Commun. **12**, Amer. Math. Soc., Providence, RI, 1997, pp. 127–148.

[46] U. Haagerup and S. Thorbjørnsen, *Random matrices with complex Gaussian entries*, preprint.

[47] U. Haagerup and S. Thorbjørnsen, *Random matrices and K-theory for exact C^*-algebras*, Doc. Math. **4** (1999), 341–450 (electronic).

[48] F. Hiai and D. Petz, *Maximizing free entropy*, Acta Math. Hungar. **80** (1998), 335–356.

[49] F. Hiai and D. Petz, *A large deviation theorem for the empirical eigenvalue distribution of random unitary matrices*, Ann. Inst. H. Poincaré Probab. Statist. **36** (2000), 71–85.

[50] F. Hiai and D. Petz, *Eigenvalue density of the Wishart matrix and large deviations*, Infin. Dimens. Anal. Quantum Probab. Relat. Top. **1** (1998), 633–646.

[51] F. Hiai and D. Petz, *Asymptotic freeness almost everywhere for random matrices*, Acta Sci. Math. (Szeged) **66** (2000), 809–834.

[52] F. Hiai and D. Petz, *Properties of free entropy related to polar decomposition*, Comm. Math. Phys. **202** (1999), 421–444.

[53] F. Hiai and D. Petz, *The Semicircle Law, Free Random Variables and Entropy*, Mathematical Surveys and Monographs, Vol. 77, Amer. Math. Soc., Providence, R.I., 2000.

[54] O. Hiwatashi, T. Kuroda, N. Nagisa, and H. Yoshida, *The free analogue of noncentral chi-square distributions and symmetric quadratic forms in free random variables*, Math. Z. **230** (1999), 63–77.

[55] O. Hiwatashi, M. Nagisa, and H. Yoshida, *The characterizations of a semicircle law by the certain freeness in a C^*-probability space*, Probab. Theory Related Fields **113** (1999), 115–133.

[56] H. Kosaki, *Free product of equivalence relations and certain amalgamated free products*, preprint.

[57] B. Krawczyk and R. Speicher, *Combinatorics of free cumulants*, J. Combin. Theory Ser. A **90** (2000), 267–292.

[58] H. van Leeuwen, *On q-deformed probability theory*, Thesis, University of Nijmegen, 1996.

[59] H. Maassen, *Addition of freely independent random variables*, J. Funct. Anal. **106** (1992), 409–438.

[60] M. Nagisa, *Stable rank of some full group C^*-algebras of groups obtained by the free product*, Internat. J. Math. **8** (1997), 375–382.

[61] M. L. Mehta, *Random Matrices*, 2nd Edition, Academic Press, Boston, 1991.

[62] A. Nica, *Asymptotically free families of random unitaries in symmetric groups*, Pacific J. Math. **157** (1993), 295–310.

[63] A. Nica, *A one-parameter family of transforms, linearizing convolution laws for probability distributions*, Comm. Math. Phys. **168** (1995), 187–207.

[64] A. Nica, *R-transform of free joint distributions and non-crossing partitions*, J. Funct. Anal. **135** (1996), 271–296.

[65] A. Nica, D. Shlyakhtenko, and R. Speicher, *Some minimization problems for the free analogue of the Fisher information*, Adv. Math. **141** (1999), 282–321.

[66] A. Nica, D. Shlyakhtenko, and R. Speicher, *Maximality of the microstates free entropy for R-diagonal elements*, Pacific J. Math. **187** (1999), 333–347.

[67] A. Nica, D. Shlyakhtenko, and R. Speicher, *R-diagonal elements and freeness with amalgamation*, preprint.

[68] A. Nica and R. Speicher, *On the multiplication of free N-tuples of noncommutative random variables*, Amer. J. Math. **118** (1996), 799–837.

[69] A. Nica and R. Speicher, *R-diagonal pairs – a common approach to Haar unitaries and circular elements*, in *Free Probability Theory*, D. V. Voiculescu (ed.), Fields Inst. Commun. **12**, Amer. Math. Soc., Providence, RI, 1997, pp. 149–188.

[70] A. Nica and R. Speicher, *A "Fourier transform" for multiplicative functions on non-crossing partitions*, J. Alg. Comb. **6** (1997), 141–160.

[71] D. Petz and F. Hiai, *Logarithmic energy as entropy functional*, in *Advances in Differential Equations and Mathematical Physics*, E. Carlen et al. (eds.), Contemp. Math. **217**, Amer. Math. Soc., Providence, RI, 1998, pp. 205–221.

[72] M. Pimsner and D. Voiculescu, *K-groups of reduced crossed products by free groups*, J. Operator Theory **8** (1982), 131–156.

[73] S. Popa, *Maximal injective subalgebras in factors associated with free groups*, Adv. Math. **50** (1983), 27–48.

[74] S. Popa, *Markov traces on universal Jones algebras and subfactors of finite index*, Invent. Math. **111** (1993), 375-405.

[75] S. Popa, *Free-independent sequences in type II_1 factors and related problems*, in *Recent Advances in Operator Algebras*, Astérisque **232** (1995), 187–202.

[76] F. Rădulescu, *The fundamental group of the von Neumann algebra of a free group with infinitely many generators is $\mathbb{R}_+ \setminus \{0\}$*, J. Amer. Math. Soc. **5** (1992), 517–532.

[77] F. Rădulescu, *A one-parameter group of automorphisms of $L(\mathbf{F}_\infty) \otimes B(H)$ scaling the trace*, C. R. Acad. Sci. Paris Sér. I Math. **314** (1992), 1027–1032.

[78] F. Rădulescu, *Stable equivalence of the weak closures of free groups convolution algebras*, Comm. Math. Phys. **156** (1993), 17–36.

[79] F. Rădulescu, *Random matrices, amalgamated free product and subfactors of the von Neumann algebra of a free group, of noninteger index*, Invent. Math. **115** (1994), 347–389.

[80] F. Rădulescu, *A type III_λ factor with core isomorphic to the von Neumann algebra of a free group, tensor $B(H)$*, in *Recent Advances in Operator Algebras*, Astérique **232** (1995), 203–209.

[81] N. Saitoh and N. Yoshida, *q-deformed Poisson random variables on q-Fock space*, J. Math. Phys. **41** (2000), 5767–5772.

[82] D. Shlyakhtenko, *Limit distributions of matrices with bosonic and fermionic entries*, in *Free Probability Theory*, D.V. Voiculescu (ed.), Fields Inst. Commun. **12**, Amer. Math. Soc., Providence, RI, 1997, pp. 241–252.

[83] D. Shlyakhtenko, *Free quasi-free states*, Pacific J. Math. **177** (1997), 329–368.

[84] D. Shlyakhtenko, *Some applications of freeness with amalgamation*, J. Reine Angew. Math. **500** (1998), 191–212.

[85] D. Shlyakhtenko, *Free entropy with respect to a completely positive map*, Amer. J. Math. **122** (2000), 45–81.

[86] R. Speicher, *Multiplicative functions on the lattice of non-crossing partitions and free convolution*, Math. Ann. **298** (1994), 611–628.

[87] R. Speicher, *Combinatorial Theory of the Free Product with Amalgamation and Operator-Valued Free Probability Theory*, Mem. Amer. Math. Soc. **627**, 1998.

[88] M. B. Ştefan, *The primality of subfactors of finite index in the interpolated free group factors*, Proc. Amer. Math. Soc. **126** (1998), 2299–2307.

[89] E. Størmer, *States and shifts on infinite free products of C^*-algebras*, in Free Probability Theory, D. V. Voiculescu (ed.), Fields Inst. Commun. **12**, Amer. Math. Soc., Providence, RI, 1997, pp. 281–291.

[90] S. Thorbjørnsen, *Mixed moments of Voiculescu's Gaussian random matrices*, J. Funct. Anal. **176** (2000), 213–246.

[91] Y. Ueda, *A minimal action of the compact quantum group $SU_q(n)$ on a full factor*, J. Math. Soc. Japan **51** (1999), 449–461.

[92] Y. Ueda, *Amalgamated free product over Cartan subalgebra*, Pacific J. Math. **191** (1999), 359–392.

[93] Y. Ueda, *Remarks on free products with respect to non-tracial states*, Math. Scand., to appear.

[94] Y. Ueda, *On the fixed-point algebra under a minimal free product type action of the quantum group $SU_q(2)$*, Internat. Math. Res. Notices (2000), 35–56.

[95] Y. Ueda and Y. Watatani, *A relation between certain interpolated Cuntz algebras and interpolated free group factors*, Proc. Amer. Math. Soc. **128** (2000), 1397–1404.

[96] D. Voiculescu, *Symmetries of some reduced free product C^*-algebras*, in *Operator Algebras and Their Connection with Topology and Ergodic Theory*, Lecture Notes in Math. **1132**, Springer-Verlag, New York, 1985, pp. 556–588.

[97] D. Voiculescu, *Addition of certain non-commuting random variables*, J. Funct. Anal. **66** (1986), 323–346.

[98] D. Voiculescu, *Multiplication of certain non-commuting random variables*, J. Operator Theory **18** (1987), 223–235.

[99] D. Voiculescu, *Circular and semicircular systems and free product factors*, in *Operator Algebras, Unitary Representations, Enveloping Algebras, and Invariant Theory*, A. Connes et al. (eds.), Birkhäuser, 1990, pp. 45–60.

[100] D. Voiculescu, *Limit laws for random matrices and free products*, Invent. Math. **104** (1991), 201–220.

[101] D. Voiculescu, *The analogues of entropy and of Fisher's information measure in free probability theory*, I, Comm. Math. Phys. **155** (1993), 71–92.

[102] D. Voiculescu, *The analogues of entropy and of Fisher's information measure in free probability theory*, II, Invent. Math. **118** (1994), 411–440.

[103] D. Voiculescu, *Operations on certain non-commutative operator-valued random variables*, in *Recent Advances in Operator Algebras*, Astérisque **232** (1995), 243–275.

[104] D. Voiculescu, *Free probability theory: Random matrices and von Neumann algebras*, in Proc. of the International Congress of Mathematicians, Vols. 1, 2 (Zürich, 1994), Birkhäuser, Basel, 1995, pp. 227–241.

[105] D. Voiculescu, *The analogues of entropy and of Fisher's information measure in free probability theory III: The absence of Cartan subalgebras*, Geometric and Functional Anal. **6** (1996), 172–199.

[106] D. Voiculescu, *The analogues of entropy and of Fisher's information measure in free probability theory, IV: Maximum entropy and freeness*, in *Free Probability Theory*, D. V. Voiculescu (ed.), Fields Inst. Commun. **12**, Amer. Math. Soc., Providence, RI, 1997, pp. 293–302.

[107] D. Voiculescu, *The analogues of entropy and of Fisher's information measure in free probability theory, V: Noncommutative Hilbert transforms*, Invent. Math. **132** (1998), 189–227.

[108] D. Voiculescu, *A strengthened asymptotic freeness result for random matrices with applications to free entropy*, Internat. Math. Res. Notices **1998**, 41–63.

[109] D. Voiculescu, *The analogues of entropy and of Fisher's information measure in free probability theory VI: Liberation and mutual free information*, Adv. Math. **146** (1999), 101–166.

[110] D. Voiculescu, *Free entropy dimension $\leqq 1$ for some generators of property T factors of type II_1*, J. Reine Angew. Math. **514** (1999), 113–118.

[111] D. V. Voiculescu, K. J. Dykema, and A. Nica, *Free Random Variables, A noncommutative probability approach to free products with applications to random matrices, operator algebras and harmonic analysis on free groups*, CRM Monograph Ser., Vol. 1, Amer. Math. Soc., Providence, RI, 1992.

[112] E. P. Wigner, *Characteristic vectors of bordered matrices with infinite dimensions*, Ann. of Math. **62** (1955), 548–564.

[113] E. P. Wigner, *On the distribution of the roots of certain symmetric matrices*, Ann. of Math. **67** (1958), 325–327.

[114] W. Woess, *Nearest neighbour random walks on free products of discrete groups*, Boll. Un. Mat. Ital. B **5** (1986), 961–982.

Graduate School of Information Sciences, Tohoku University, Aoba-ku, Sendai 980-8577, Japan

Translated by FUMIO HIAI

Amer. Math. Soc. Transl.
(2) Vol. **204**, 2001

Dynamics of Kleinian Groups —the Hausdorff Dimension of Limit Sets

Katsuhiko Matsuzaki

1. Introduction

A Kleinian group is a discrete group of fractional linear transformations of the complex plane. In the mathematics of the 19th century, particular discrete groups such as Fuchsian groups began attracting attention in the theory of ordinary differential equations or, even before, in the theory of modular functions. At the end of the 19th century, Fricke and Klein (see [**Mg**]) created a general theory of discrete groups of fractional linear transformations and obtained some fundamental results such as complexity of the limit set (the set of accumulation points of the orbit) in an era when the word "fractal" had not yet been born. On the other hand, Poincaré [**Po**] extended the concept of Fuchsian groups to three dimensions and regarded a Kleinian group as a discrete group of isometries of the 3-dimensional hyperbolic space (although there had not been much progress from this point of view for more than half a century).

Classical 2-dimensional theories of Kleinian groups after Fricke and Klein developed gradually in the first half of the 20th century. At the same time, iteration of rational functions was studied by Fatou and Julia around 1920, and some analogies to Kleinian groups were recognized. Then these brothers of complex dynamics had spent their pupal stage, and the time was ripe for the next. The first metamorphosis of the theory of Kleinian groups was brought about by the completion of fundamental theorems on quasiconformal maps after 1960. By virtue of this, in addition to the conventional analysis of the phase space of dynamics, giving deformation on the dynamics became a powerful tool, and it developed further to the analysis of the parameter space itself, namely, to a branch of the theory of Teichmüller spaces. Ahlfors and Bers established the foundation of complex analytic methods for the study of Kleinian groups.

This article originally appeared in Japanese in Sūgaku **51** (2) (1999), 142–160.

2000 *Mathematics Subject Classification.* Primary 57S30; Secondary 30F40.

Added in translation. The following paper by the author is related to the arguments in Section 6 of the present article: K. Matsuzaki, *Convergence of the Hausdorff dimension of the limit sets of Kleinian groups*, in *In the Tradition of Ahlfors and Bers: Proceedings of the First Ahlfors-Bers Colloquium* (Stony Brook, NY, 1998), Contemporary Math., vol. 256, Amer. Math. Soc., Providence, RI, 2000, pp. 243–254.

However, the real nature of Kleinian groups becomes clearer only when we observe them as isometry groups of the 3-dimensional hyperbolic space. As 3-dimensional topology developed, the recurrence to Poincaré's viewpoint had appeared in the work of Marden and Maskit in the 1970s, and at last the theory of 3-dimensional hyperbolic manifolds due to Thurston [**Th**] brought a revolutionary turning point to Kleinian groups and created a big current of mathematics which has lasted until now. Moreover, several fields of mathematics have been involved with this current and, as a consequence, the range of the theory of Kleinian groups has been expanded.

Among these modern theories of Kleinian groups, another event which seems no less important was Sullivan's work—so to speak, the renaissance of classical analysis and the unification of complex dynamics. He generalized the Hopf-Tsuji ergodicity theorem, the Mostow rigidity theorem and the Ahlfors finiteness theorem by dynamical methods [**S2**], and imported quasiconformal maps into the study of the iteration of rational maps, and tried to make a unified interpretation with Kleinian groups. Also, he utilized a measure on the limit set which was introduced by Patterson to clarify how the Hausdorff dimension, which had until then been a primitive index measuring complexity of the limit set as a fractal, reflected geometric features of the hyperbolic manifold well [**S1**].

This expository paper concentrates on topics on the Hausdorff dimension of limit sets among the complex dynamical theories of Kleinian groups. Sullivan's theories appeared in the 1980s, and since then a lot of good expositions have been published, among which the monograph by Nicholls [**N**] is popular; in it the reader can find detailed proofs of several propositions in this paper, even if they are not specifically referred to. However, certain results in recent papers (in particular, a paper by Bishop and Jones [**BJ**]) succeeded in simplifying and completing Sullivan's original work. In addition, a satisfactory answer was given to a problem asking about the change of the Hausdorff dimension of the limit sets under deformation of Kleinian groups, a typical problem on dynamics (McMullen [**Mc**]). In this paper, including these topics, we try to reorganize important results and to explain them in such a way as seems most simple and clear (for the author). Basic terminology and facts on Kleinian groups will be stated minimally when they are necessary; when they are omitted, we can consult monographs on Kleinian groups, say [**MT**].

2. The limit sets of Kleinian groups

A conformal automorphism of the Riemann sphere $\hat{\mathbf{C}}$ is a fractional linear transformation (an orientation-preserving Möbius transformation), and the group of all these transformations is identified with $\mathrm{PSL}_2(\mathbf{C}) = \mathrm{SL}_2(\mathbf{C})/\pm I$ as a Lie group. A Kleinian group is a discrete subgroup of this Lie group. In this paper we always assume that *Kleinian groups have no elements of finite order other than the identity.* Since a Möbius transformation is originally defined as a composition of reflections with respect to circles or lines on the complex plane, its action is naturally extended to the upper half-space

$$H^3 = \{(x, y, t) \mid t > 0\}$$

(the Poincaré extension). Moreover, if we provide the hyperbolic metric

$$d\rho^2 = (dx^2 + dy^2 + dt^2)/t^2$$

for H^3 and regard it as a model of the 3-dimensional hyperbolic space, Möbius transformations are isometric automorphisms of (H^3, ρ). Furthermore, mapping the upper half-space to the unit ball by the Cayley transformation, we obtain the unit ball model (B^3, ρ) of the hyperbolic space. In this model, the boundary S^2 of B^3 is the sphere located at infinity of the hyperbolic space.

A Kleinian group Γ acts freely and properly discontinuously on the hyperbolic space as a group of orientation-preserving isometries, and the quotient space $N_\Gamma = B^3/\Gamma$ is a hyperbolic manifold. On the other hand, if we consider the maximal open subset $\Omega(\Gamma)$ (which is called the region of discontinuity) of S^2 where Γ acts properly discontinuously, its quotient space $\Omega(\Gamma)/\Gamma$ is a complex manifold (Riemann surface). We assume the manifold $(B^3 \cup \Omega(\Gamma))/\Gamma$ with boundary to have the hyperbolic structure inside and to have the complex structure on the boundary, and we call it a Kleinian manifold.

The complement of the region of discontinuity $\Omega(\Gamma)$ of S^2 is called the limit set and denoted by $\Lambda(\Gamma)$. This can be alternatively defined as follows.

DEFINITION (Limit set). Let $\Gamma(0)$ be the orbit of the origin $0 \in B^3$ under the action of a Kleinian group Γ. Then the set of accumulation points of $\Gamma(0)$ in the Euclidean topology, which is on S^2, is defined as the limit set of Γ and denoted by $\Lambda(\Gamma)$.

If the limit set $\Lambda(\Gamma)$ contains more than two points, then it is an uncountable perfect set consisting of infinitely many points. Otherwise Γ is said to be elementary. In this paper we always assume that *Kleinian groups are nonelementary.* Another characterization of the limit set is that it is the minimal, nonempty, Γ-invariant, closed subset of S^2. We consider the closed convex hull of the union of all geodesic lines with end points in $\Lambda(\Gamma)$, and take the quotient by Γ, which is a subset of N_Γ. This is called the *convex core* and denoted by C_Γ. The convex core C_Γ is the minimal, convex, closed subset of N_Γ such that the inclusion map into N_Γ is a homotopy equivalence.

As a generalization of finiteness of the volume of N_Γ, we define the following property of N_Γ, which means "the finiteness on the limit set".

DEFINITION (Geometric finiteness). We say that a Kleinian group Γ (or a hyperbolic manifold N_Γ) is geometrically finite if Γ is finitely generated and the volume of the convex core C_Γ is finite.

For a geometrically finite Kleinian group Γ, the limit set is decomposed into

$$\Lambda(\Gamma) = \Lambda_c(\Gamma) \cup \Lambda_p(\Gamma).$$

Here $\Lambda_c(\Gamma)$ is the *conical limit set* of Γ, meaning that $x \in S^2$ belongs to $\Lambda_c(\Gamma)$ if $\Gamma(0)$ accumulates on x within a bounded distance of a geodesic ray towards x. On the other hand, $\Lambda_p(\Gamma)$ is the set of all parabolic fixed points of Γ, which is a countable set. In fact, if G is geometrically finite, then any point of $\Lambda_p(\Gamma)$ is a *bounded parabolic fixed point.*

DEFINITION (Hausdorff measure, dimension). For a subset E of S^2, we take a covering $E \subset \bigcup \Delta(x_i, r_i)$ by countably many disks $\Delta(x_i, r_i)$ with respect to the spherical metric, where the radius r_i is less than $\delta > 0$. For an arbitrary real number $s \geq 0$, we set $\mathcal{H}_s^\delta(E) = \inf \sum r_i^s$, where the infimum is taken over all such coverings of E. Next we take the limit

$$\mathcal{H}_s(E) = \lim_{\delta \to 0} \mathcal{H}_s^\delta(E) = \sup_{\delta > 0} \mathcal{H}_s^\delta(E)$$

as $\delta \to 0$. Then $\mathcal{H}_s$ is an outer measure on S^2, and the restriction to $\mathcal{H}_s$-measurable sets (which includes the Borel sets) is, by definition, the s-dimensional Hausdorff measure.

For any set $E \subset S^2$, if $s < t$, then $\mathcal{H}_s^\delta(E) \geq \delta^{s-t}\mathcal{H}_t^\delta(E)$. Hence there exists a unique number d such that if $0 \leq s < d$, then $\mathcal{H}_s(E) = \infty$, and if $d < s < \infty$, then $\mathcal{H}_s(E) = 0$. This number d is, by definition, the Hausdorff dimension of E and is denoted by $\dim E$.

There are detailed discussions of the Hausdorff measure in the monographs [**Fa**], [**Fe**] and [**Ro**].

Ahlfors proved that, for a geometrically finite Kleinian group Γ, the 2-dimensional Hausdorff measure satisfies $\mathcal{H}_2(\Lambda(\Gamma)) = 0$ if $\Lambda(\Gamma) \neq S^2$ (see Corollary 14). The statement that any finitely generated Kleinian group should satisfy this conclusion is called the Ahlfors conjecture, and it is not yet completely solved. It is known that this follows from the Marden conjecture (see [**O**]) asserting that N_Γ is *topologically tame*, namely that N_Γ is homeomorphic to the interior of some compact manifold with boundary for any finitely generated Kleinian group Γ [**C2**].

3. Critical exponent of convergence

We define the other classical index for Kleinian groups as follows:

Definition (Critical exponent). For a Kleinian group Γ, we define

$$\delta(\Gamma) = \inf\{s \geq 0 \mid g_s := \sum_{\gamma \in \Gamma} \exp(-s\rho(0, \gamma(0))) < \infty\}$$

to be the critical exponent for Γ, where ρ denotes the hyperbolic distance. Here g_s is called the s-dimensional *Poincaré series*. If the $\delta(\Gamma)$-dimensional Poincaré series diverges, then Γ is said to be of *divergence type*, and if it converges, then it is of *convergence type*.

In the unit ball model B^3, since $\exp(-\rho(0, \gamma(0)))$ is comparable with $1 - |\gamma(0)|$, the critical exponent can be defined by the convergence or divergence of $\sum_{\gamma \in \Gamma}(1 - |\gamma(0)|)^s$. In addition, by using the equation

$$|\gamma'(x)| = \frac{1 + |\gamma(0)|}{|x - \gamma^{-1}(0)|^2}(1 - |\gamma(0)|)$$

for $x \in S^2$, the critical exponent can also be defined by the convergence or divergence of $\sum_{\gamma \in \Gamma} |\gamma'(x)|^s$ ($x \in \Omega(\Gamma)$), where $|\gamma'(x)|$ means the magnifying rate of a conformal map γ of $\mathbf{R}^3$ at x.

Let $n(r)$ denote the number of points of $\Gamma(0)$ within a distance r of the origin. Then it can be proved that

$$\delta(\Gamma) = \limsup_{r \to \infty} \frac{\log n(r)}{r}.$$

In particular, this implies that $\delta(\Gamma) \leq 2$. Also $\delta(\Gamma) > 0$ can be proved in several ways.

Concerning the relationship between the critical exponent and the Hausdorff dimension, the following theorem due to Bishop and Jones [**BJ**] is the latest word.

Theorem 1. *Any Kleinian group* Γ *satisfies* $\dim \Lambda_c(\Gamma) = \delta(\Gamma)$.

First we show the easier estimate from above: $\dim \Lambda_c(\Gamma) \le \delta(\Gamma)$. For any $\gamma \in \Gamma$, let $b(\gamma, t)$ denote the "*shadow*" on S^2, cast by the rays from the origin, of a ball with radius $t > 0$ and center $\gamma(0)$. Then, numbering all elements of Γ as $\{\gamma_n\}_{n=1}^{\infty}$, we can represent the conical limit set as

$$\Lambda_c(\Gamma) = \bigcup_{t>0} \Lambda_c^t(\Gamma), \quad \Lambda_c^t(\Gamma) := \bigcap_{N \ge 1} \bigcup_{n \ge N} b(\gamma_n, t).$$

PROPOSITION 2. *If the s-dimensional Poincaré series converges, then the s-dimensional Hausdorff measure of $\Lambda_c(\Gamma)$ is zero.*

PROOF. It suffices to prove that the s-dimensional Hausdorff measure of $\Lambda_c^t(\Gamma)$ is zero for each fixed $t > 0$. This follows from the fact that the radius of the shadow $b(\gamma_n, t)$ is comparable with $1 - |\gamma_n(0)|$. □

The assertion of Theorem 1 was already known for Fuchsian groups in general and for geometrically finite Kleinian groups [**S1**], [**S4**]. The value of Bishop and Jones' result lies in that they proved it for Kleinian groups in general. We will introduce an interpretation of the original proof by McMullen, who utilizes *quasi-geodesics*. Here we say that a piecewise geodesic curve in the hyperbolic space is an (L, θ)-quasigeodesic if the length of each geodesic segment is greater than $L > 0$ and the angle between any two consecutive segments is greater than $\theta > 0$.

Now we will prove the other inequality, $\dim \Lambda_c(\Gamma) \ge \delta(\Gamma)$. For any $\epsilon > 0$, set $s' = \delta(\Gamma) - \epsilon$ and $s = \delta(\Gamma) - 2\epsilon$. Then the s'-dimensional Poincaré series diverges. We will prove that the s-dimensional Hausdorff measure of $\Lambda_c(\Gamma)$ is not zero.

We construct a tree T rooted at the origin in B^3 as follows: The vertices of T are in a subset of the orbit $\Gamma(0)$; each edge of T is a geodesic segment of length greater than $L > 0$; and the angle between two edges at each vertex is greater than $\theta > 0$. Then, by a property of quasigeodesics, a piecewise geodesic curve from a vertex of T to a descendent vertex along the edges lies within a constant distance of the geodesic line connecting the two vertices, where the constant depends only on L and θ. From this fact, we can see that a set $E(T) \subset S^2$ of the ends of T is contained in $\Lambda_c(\Gamma)$. Furthermore, we require the following additional properties for T: the sum taken over all vertices v' that are direct children of any vertex v in T satisfies

$$(*) \qquad \sum_{v'} \exp(-s\rho(v', 0)) \ge \exp(-s\rho(v, 0)),$$

and the length of each edge e is bounded by a constant L'.

Assuming the existence of such a tree T, we prove that $\mathcal{H}_s(E(T)) > 0$. We take an arbitrary covering $\bigcup \Delta(x_i, r_i)$ of $E(T)$ by disks. For each path in T from the origin towards an end, if it terminates at an end contained in a disk $\Delta(x_i, r_i)$, we cut the path at the vertex where it first enters into the hemiball B_i in B^3 spanning $\Delta(x_i, 2r_i)$ and take off the far part from there. In this manner, we construct a finite tree T'. Then the sum taken over all vertices v' on the boundary of T' satisfies

$$\sum \exp(-s\rho(v', 0)) \ge \exp(-s\rho(0, 0)) = 1$$

by $(*)$. Since $\exp(-s\rho(v', 0))$ and $(1 - |v'|)^s$ are comparable, there exists a constant $C > 0$ such that if $v' \in B_i$, then $C(2r_i)^s \ge \exp(-s\rho(v', 0))$. The number of $v' \in \partial T'$ that are contained in a hemiball B_i is bounded by a constant K that depends only on $a = \inf_{\gamma \in \Gamma - \{\mathrm{id}\}} \rho(\gamma(0), 0)$ and on L'. Hence

$$2^s CK \sum_i r_i^s \geq \sum_{v' \in \partial T'} \exp(-s\rho(v', 0)) \geq 1.$$

This implies that, for any covering of $E(T)$ by disks, the sum of their radii to the s-th power is not less than $(2^s CK)^{-1}$, and thus we obtain

$$\mathcal{H}_s(\Lambda_c(\Gamma)) \geq \mathcal{H}_s(E(T)) \geq (2^s CK)^{-1} > 0.$$

Next we give a method of constructing a tree T that satisfies the above requirements. Let A_1 be a cone consisting of the rays from the origin towards the points in a disk Δ_1 on S^2. We can choose Δ_1 so that a partial sum of the s'-dimensional Poincaré series $g_{s'}$ taken over the points of $\Gamma(0)$ that are contained in A_1 diverges. Moreover, since Γ is nonelementary, we can take another disk Δ_2 on S^2 disjoint from Δ_1 so that the partial sum of $g_{s'}$ taken over the points in the corresponding cone A_2 to Δ_2 also diverges. Let θ be half of the minimum of the angles between a ray from the origin in A_1 and one in A_2. For a fixed $l > 0$, we choose $L > 0$ so that any two (L, θ)-quasigeodesics that connect the same points in B^3 lie within distance l of each other.

We divide the cone A_i $(i = 1, 2)$ according to the hyperbolic distance from the origin and let $A_{i,n}$ $(n = 1, 2, \dots)$ be the block in A_i where the distance is in the interval $[n, n+1)$. We consider a partial sum of the Poincaré series $g_{s'}$ taken over the points of $\Gamma(0)$ in $A_{i,n}$. Since $g_{s'}$ diverges, for any $M > 0$ there exist infinitely many blocks $\{A_{i,n(k)}\}_{k=1,2,\dots}$ such that the partial sum of g_s taken over $A_{i,n(k)}$ is greater than M. On the other hand, we can find a way of dividing $\Gamma(0) \cap A_{i,n}$ into b groups, where b is an integer depending only on a and l, such that, if two (L, θ)-quasigeodesic rays from the origin pass through distinct points in the same group, then they are at least l distant from each other in $A_{i,n}$. We take M sufficiently larger than b and choose a block $A_{i,n(k)}$ with $n(k) \geq L$. At least one group among the b groups is such that the partial sum of g_s taken over the group is greater than $M/b >> 1$.

We adopt all points in the group as the vertices $\{v\}$ in the first generation of the tree and connect the origin and each vertex v by a geodesic segment, which is an edge e. The way of choosing the vertices $\{v'\}$ in the second generation, which are children of a vertex v, is as follows: We regard $v = \gamma(0)$ as the new origin and take two cones $\gamma(A_1)$ and $\gamma(A_2)$ with the vertex v. Then the angle between the edge e and at least one of the two cones is not less than θ. Choose certain points of $\Gamma(0)$ in the blocks of the one cone in exactly the same manner as before and adopt them as the vertices $\{v'\}$. Connect v and each v' by a geodesic edge e'.

Repeating this process, we have a family of (L, θ)-quasigeodesic rays starting from the origin, branching at the vertices and extending to infinity. This family of quasigeodesic rays forms a tree because, once they branch, they never meet again due to our construction. For each vertex v and its children $\{v'\}$, we have

$$\sum_{v'} \exp(-s\rho(v', 0)) \approx \sum_{v'} \exp(-s\rho(v', v)) \cdot \exp(-s\rho(v, 0))$$

because the quasigeodesics satisfy $\rho(v', 0) \approx \rho(v', v) + \rho(v, 0)$. Then, since

$$\sum_{v'} \exp(-s\rho(v', v)) >> 1$$

by construction, we obtain the inequality $(*)$.

4. Measures on the limit set

We consider measures on the limit set that are invariant under the group action (in the sense of the following definition). The Hausdorff measure is invariant under Möbius transformations. There is a standard construction of an invariant probability measure with support on the limit set, which reflects the distribution of the orbit of a Kleinian group. Using this measure, we can estimate the Hausdorff measure and dimension of the limit set.

DEFINITION (Invariant measure). For a Kleinian group Γ, a Borel measure μ on S^2 is said to be an s-dimensional, Γ-invariant measure if

$$\mu(\gamma E) = \int_E |\gamma'|^s d\mu$$

for any Borel measurable set E on S^2 and for any $\gamma \in \Gamma$.

We consider the range of dimensions where a Γ-invariant probability measure exists. We define the infimum of such dimensions (which is actually the minimum by taking the weak limit of a sequence of measures) as the *critical dimension* and denote it by $\alpha(\Gamma)$.

Patterson [**Pt1**] and Sullivan [**S1**] constructed a $\delta(\Gamma)$-dimensional, Γ-invariant probability measure as follows: For $s > \delta(\Gamma)$, letting each term of the convergent Poincaré series g_s be a weight of the Dirac measure $\delta_{\gamma(0)}$, we consider the sum

$$\sum_{\gamma \in \Gamma} \exp(-s\rho(0, \gamma(0)))\delta_{\gamma(0)}.$$

Dividing it by the total mass, we have a probability measure on $\overline{B^3}$. Then, letting $s \downarrow \delta(\Gamma)$, we have a weak limit of a subsequence of the measures. We can prove that the probability measure constructed in this way has the support on the limit set and that it has $\delta(\Gamma)$-dimensional Γ-invariance. (Precisely speaking, this is true only in the case where Γ is of divergence type. If Γ is of convergence type, we have to adjust the given weight so that the weak limit measure has no support inside of B^3.)

LEMMA 3. *For any Kleinian group Γ, there exists a $\delta(\Gamma)$-dimensional, Γ-invariant probability measure. Hence $\alpha(\Gamma) \leq \delta(\Gamma)$.*

Next, we investigate the relationship between an s-dimensional, Γ-invariant probability measure μ_s and the s-dimensional Hausdorff measure. To this end, the following *shadow lemma* due to Sullivan [**S1**] is crucial; it asserts that the mass of the shadow $b(\gamma, t)$ (defined in Section 3) measured by μ_s is comparable with the radius $r(b(\gamma, t))$ of the shadow to the s-th power. Consult [**Pt3**] for its proof.

LEMMA 4. *Let μ_s be an s-dimensional, Γ-invariant probability measure. Then, for a sufficiently large t, there exists a constant $A > 0$ such that*

$$\frac{1}{A}\mu_s(b(\gamma, t)) < r(b(\gamma, t))^s < A\mu_s(b(\gamma, t))$$

for all but finitely many $\gamma \in \Gamma$.

Applying this lemma directly to the proof of Proposition 2 and interpreting the conclusion as a statement on a Γ-invariant probability measure, we have the following.

PROPOSITION 5. *If the s-dimensional Poincaré series converges, then $\Lambda_c(\Gamma)$ is a null set for any s-dimensional, Γ-invariant measure μ_s.*

The restriction of the s-dimensional Hausdorff measure $\mathcal{H}_s$ to the conical limit set $\Lambda_c(\Gamma)$ is denoted by h_s. This is an s-dimensional, Γ-invariant measure (including the case where the total mass is zero or infinity). In order to estimate h_s from above in terms of μ_s, we use Lemma 4 as well as the fact that we can choose a covering of $\Lambda_c(\Gamma)$ by shadows that have no intersection (or few intersections) with each other. In general, this kind of statement is called the *covering theorem* and, for example, the Vitali covering theorem (see [**Fa**]) is applicable to our case.

LEMMA 6. *If an s-dimensional, Γ-invariant probability measure μ_s exists, then there is a constant $A' < \infty$ such that $h_s \leq A'\mu_s$. Hence* $\dim \Lambda_c(\Gamma) \leq s$ *and thus* $\dim \Lambda_c(\Gamma) \leq \alpha(\Gamma)$.

By Lemmas 3 and 6, the above three values $\dim \Lambda_c$, α and δ are in this order from low to high. Then, applying Theorem 1, we see that they are actually coincident with each other.

THEOREM 7. *Any Kleinian group Γ satisfies* $\dim \Lambda_c(\Gamma) = \alpha(\Gamma) = \delta(\Gamma)$.

We consider the following more detailed conditions on convergence of Poincaré series and on nullity of the conical limit set at the critical dimension $\delta = \delta(\Gamma)$.

(1) The δ-dimensional Poincaré series g_δ converges.
(2) Some/any δ-dimensional, Γ-invariant probability measure μ_δ satisfies $\mu_\delta(\Lambda_c(\Gamma)) = 0$.
(3) The δ-dimensional Hausdorff measure $\mathcal{H}_\delta$ satisfies $\mathcal{H}_\delta(\Lambda_c(\Gamma)) = 0$.

Although only the implication (1) $\Rightarrow$ (2) was proved in Proposition 5, the converse (2) $\Rightarrow$ (1) is also true. To see this, we take the unit tangent bundle $T_1 N_\Gamma$ of the hyperbolic manifold N_Γ and introduce a measure m on $T_1 N_\Gamma$ from the product of the hyperbolic volume element and a Γ-invariant probability measure μ. Then we consider a condition that the *geodesic flow* $g(v,t) : T_1 N_\Gamma \times \mathbf{R} \to T_1 N_\Gamma$ is conservative with respect to the measure m. In other words, for almost all points $v \in T_1 N_\Gamma$ with respect to m, there exists an infinite sequence $t_n \to \infty$ such that $g(v, t_n)$ belong to a compact subset of $T_1 N_\Gamma$. Clearly this is equivalent to the condition $\mu(\Lambda_c(\Gamma)) = 1$. Sullivan proved that, when μ is n-dimensional, if the Poincaré series g_n diverges then the geodesic flow is conservative [**S2**], and that this is extendable to any dimension [**S3**], [**N**]. Further, Thurston gave an elementary proof directly (without using a geodesic flow) of the fact that if g_n diverges, then $\Lambda_c(\Gamma)$ has full n-dimensional measure (see [**Ah**], [**N**]). Tukia [**Tu2**] generalized this method to any dimension.

THEOREM 8. *If Γ is of divergence type, then $\Lambda_c(\Gamma)$ is positive (actually full) with respect to any $\delta(\Gamma)$-dimensional, Γ-invariant probability measure.*

Next we show that a $\delta(\Gamma)$-dimensional, Γ-invariant probability measure is unique for a Kleinian group Γ of divergence type. The uniqueness is equivalent to the *ergodicity* of the group action.

PROPOSITION 9. *Let μ_s be an s-dimensional, Γ-invariant probability measure. Then there exists no other s-dimensional, Γ-invariant probability measure that is absolutely continuous to μ_s if and only if it satisfies the following ergodic condition: any Γ-invariant measurable set A on S^2 satisfies either $\mu_s(A) = 0$ or $\mu_s(S^2 - A) = 0$.*

In general, as the following lemma shows, any Γ-invariant probability measure μ satisfies the ergodic condition on the conical limit set $\Lambda_c(\Gamma)$. Hence, if $\Lambda_c(\Gamma)$ is positive with respect to μ, which implies that μ must be $\delta(\Gamma)$-dimensional, then it is unique.

LEMMA 10. *If a Γ-invariant measurable subset A of $\Lambda_c(\Gamma)$ satisfies $\mu(A) > 0$ for an s-dimensional, Γ-invariant probability measure μ, then $\mu(A) = 1$.*

PROOF. We take a density point x of A and choose a sequence $\gamma_n(0)$ out of the orbit that converges to x conically. Passing to a subsequence, we may assume that $\gamma_n^{-1}(0) \to y$ for some $y \in S^2$. Since x is a density point, a sequence of shadows $b(\gamma_n, t)$ satisfies

$$\lim_{n\to\infty} \frac{\mu(b(\gamma_n, t) - A)}{\mu(b(\gamma_n, t))} \to 0.$$

We map the shadow $b(\gamma_n, t)$ by γ_n^{-1} and consider the ratio of A in the image. Estimating $|(\gamma_n^{-1})'|$, we have a constant C depending only on t such that

$$\frac{\mu(\gamma_n^{-1} b(\gamma_n, t) - A)}{\mu(\gamma_n^{-1} b(\gamma_n, t))} = \frac{\int_{b(\gamma_n,t)-A} |(\gamma_n^{-1})'|^s d\mu}{\int_{b(\gamma_n,t)} |(\gamma_n^{-1})'|^s d\mu} \le C \frac{\mu(b(\gamma_n, t) - A)}{\mu(b(\gamma_n, t))} \quad (\to 0).$$

Then, for any $\epsilon > 0$, there exists some t such that $\mu(\gamma_n^{-1} b(\gamma_n, t)) > 1 - \lambda - \epsilon$ for all sufficiently large n, where $\mu(\{y\}) = \lambda$. Hence, when $y \in A$,

$$\frac{\mu(A \cap \gamma_n^{-1} b(\gamma_n, t))}{\mu(\gamma_n^{-1} b(\gamma_n, t))} < \frac{\mu(A) - \lambda}{1 - \lambda - \epsilon}.$$

Since the left-hand side converges to 1 as $n \to \infty$, we see that $1 - \lambda - \epsilon \le \mu(A) - \lambda$. Thus we have $\mu(A) = 1$, for ϵ is arbitrary. If $y \notin A$, we have $1 - \lambda \le \mu(A)$ in the same way. However, this is possible only when $\lambda = 0$. □

DEFINITION (The Patterson–Sullivan measure). For a Kleinian group Γ of divergence type, the unique $\delta(\Gamma)$-dimensional, Γ-invariant probability measure is called the *Patterson–Sullivan measure*, or, for short, the PS measure. The total mass of the PS measure is on the conical limit set.

Finally we consider the implication (3) $\Rightarrow$ (2). Note that the converse is true, which can be seen from Lemma 6 or alternatively from Proposition 2 if we use the equivalence of (1) and (2). In the definition of the Hausdorff measure, all the disks are candidates for a covering of a given set; however, we restrict the disks to the family $\mathcal{F}_t = \{b(\gamma, t)\}_{\gamma\in\Gamma}$ of shadows of balls of radius t with center at the orbit and define a new Hausdorff measure $\mathcal{H}_s(\cdot\,; \mathcal{F}_t)$ in the same way. Then, by the shadow lemma and by the covering theorem, we can see that $\Lambda_c^t(\Gamma)$ is a null set for an s-dimensional, Γ-invariant probability measure μ_s if and only if it is a null set for $\mathcal{H}_s(\cdot\,; \mathcal{F}_t)$. Hence it remains to compare $\mathcal{H}_s(\cdot\,; \mathcal{F}_t)$ and $\mathcal{H}_s(\cdot)$. The inequality $\mathcal{H}_s(\cdot) \le \mathcal{H}_s(\cdot\,; \mathcal{F}_t)$ is clear by definition; the converse is the problem. If $\Lambda_c^t(\Gamma)$ is a null set for $\mathcal{H}_s(\cdot)$, then it is also one for $\mathcal{H}_s(\cdot\,; \mathcal{F}_t)$. However, this is not true in general.

REMARK. Assume that Γ is a geometrically finite Kleinian group without a parabolic element. Then, for any disk $\Delta(x, r)$ with sufficiently small radius that contains a point of $\Lambda_c^t(\Gamma)$, there exists a shadow $b(\gamma, t) \subset \Delta(x, 2r)$ such that if the radius is multiplied by a uniform constant, then the resulting shadow contains

$\Delta(x, 2r)$. Hence, in this case, both Hausdorff measures are comparable on $\Lambda_c^t(\Gamma)$ and thus (3) $\Rightarrow$ (2). However, even if Γ is geometrically finite, this is not true in certain cases where Γ contains a parabolic element. Sullivan [**S4**] gave a condition for this implication to be satisfied in terms of the inequalities between $\delta(\Gamma)$ and the maximal rank of parabolic subgroups of Γ.

We have seen that, for any Kleinian group Γ, there exists a $\delta(\Gamma)$-dimensional, Γ-invariant probability measure, and if in addition Γ is of divergence type, then it is unique, that is, it is the PS measure. Although the uniqueness is not satisfied in general, a measure that is constructed in a standard manner has the following property [**S4**].

THEOREM 11. *For any Kleinian group Γ, there exists a $\delta(\Gamma)$-dimensional, Γ-invariant probability measure μ whose support is on the limit set and which has no atom (point mass) on bounded parabolic fixed points.*

Indeed, since the measure μ constructed in Lemma 3 has its support on the limit set, it suffices only to prove that μ has no atom (point mass) on bounded parabolic fixed points. The PS measure automatically satisfies this condition.

DEFINITION (Change of base points). For an s-dimensional, Γ-invariant probability measure μ and a point $x \in B^3$, we define a measure μ_x that is absolutely continuous with respect to μ and vice versa, as

$$\mu_x(E) = \int_E P(x, \zeta)^s d\mu(\zeta), \quad P(x, \zeta) = \frac{1 - |x|^2}{|x - \zeta|^2}.$$

Here $P(x, \zeta)$ is the Poisson kernel, which is the magnifying rate $|h'_x(\zeta)|$ at $\zeta \in S^2$ of a Möbius transformation h_x of B^3 that maps x to the origin 0. Then the Radon–Nikodým derivative is

$$\frac{d\mu_x}{d\mu_y}(\zeta) = \left(\frac{P(x, \zeta)}{P(y, \zeta)}\right)^s.$$

We consider a Γ-invariant function $\varphi_\mu(x) = \mu_x(S^2)$, where the total mass of μ_x is regarded as a function of $x \in B^3$. We will see that the growth order of $\varphi_\mu(x)$ as x tends conically to ξ depends on whether ξ is an atom for μ or not.

In general, when $\xi \in S^2$ is an atom for an s-dimensional, Γ-invariant probability measure μ, we can see from an estimate of the Poisson kernel that, for a sequence of points $x \in B^3$ that converges conically to ξ, there exists a constant $C_1 > 0$ such that

$$\varphi_\mu(x) \geq \mu_x(\{\xi\}) = P(x, \xi)^s \mu(\{\xi\}) \geq C_1 e^{s\rho(0,x)}.$$

This, in particular, implies that a conical limit point ξ of Γ cannot be an atom for μ. Contrary to this case, when ξ is a bounded parabolic fixed point of Γ and it is not an atom, we have the following result. For simplicity of notation, ξ is regarded as a bounded parabolic fixed point of rank 0 if $\xi \in \Omega(\Gamma)$.

LEMMA 12. *Let μ be an s-dimensional, Γ-invariant probability measure with support on $\Lambda(\Gamma)$. Assume that ξ is a bounded parabolic fixed point of Γ of rank k $(k = 0, 1, 2)$ that is not an atom for μ. Then, for a sequence $x \in B^3$ that converges conically to ξ, there exists a constant $C_2 < \infty$ that satisfies*

$$\varphi_\mu(x) \leq C_2 e^{(k-s)\rho(0,x)}.$$

PROOF. Using the upper half-space model H^3, we may assume that $\xi = \infty$ and that μ_x is a measure on $\mathbf{C}$. Setting $P(x,\zeta) = 2t/|x-\zeta|^2$ for $x = (z,t) \in H^3$ (which is the magnifying rate $|h'_x(\zeta)|$ of a Möbius transformation $h_x : H^3 \to B^3$ that maps x to 0), we have

$$\frac{d\mu_x}{d\mu_{x_0}}(\zeta) = \left(\frac{P(x,\zeta)}{P(x_0,\zeta)}\right)^s = \left(\frac{t(1+|\zeta|^2)}{t^2+|\zeta-z|^2}\right)^s,$$

where $x_0 = (0,1)$. Let J be the stabilizer of ∞ in Γ and E a fundamental domain for J on $\mathbf{C}$. Then

$$\begin{aligned}\mu_x(\mathbf{C}) = \sum_{j\in J}\mu_x(j(E)) &= \sum_{j\in J}\mu_{j^{-1}(x)}(E)\\ &= \sum_{j\in J}\int_E \frac{d\mu_{j^{-1}(x)}}{d\mu_{x_0}}(\zeta)d\mu_{x_0}(\zeta)\\ &= t^s\sum_{j\in J}\int_E\left(\frac{1+|\zeta|^2}{t^2+|\zeta-j^{-1}(z)|^2}\right)^s d\mu_{x_0}(\zeta).\end{aligned}$$

Since the support $\Lambda(\Gamma)$ of μ_{x_0} restricted to E is bounded, the sum of the integrals in the last term is comparable with

$$\int_{\mathbf{R}^k}\frac{dx_1\cdots dx_k}{(t^2+x_1^2+\cdots+x_k^2)^s} = t^{k-2s}\int_{\mathbf{R}^k}\frac{dx_1\cdots dx_k}{(1+x_1^2+\cdots+x_k^2)^s}.$$

(This, in particular, implies that $s > k/2$ for $k = 1,2$.) We transfer these things to B^3. When x converges to ξ conically, t is comparable with $e^{\rho(x_0,x)}$. Hence we can take a required constant C_2 as in the assertion. □

PROOF OF THEOREM 11. We take a $\delta(\Gamma)$-dimensional, Γ-invariant probability measure μ with support on $\Lambda(\Gamma)$, which was constructed in Lemma 3. Suppose that Γ has a bounded parabolic fixed point ξ of rank $k = 1,2$. Then $\delta(\Gamma) > k/2$, which can be seen from the proof of Lemma 12 above, or more simply from Lemma 30, below. For a sequence $x \in B^3$ that converges conically to ξ, whether ξ is an atom or not makes a difference of growth order:

$$\begin{aligned}\varphi_\mu(x) &\geq C_1e^{\delta(\Gamma)\rho(0,x)};\\ \varphi_\mu(x) &\leq C_2e^{(k-\delta(\Gamma))\rho(0,x)} = C_2e^{(\delta(\Gamma)-\epsilon)\rho(0,x)} \quad (\epsilon = 2\delta(\Gamma)-k>0).\end{aligned}$$

The measure μ was constructed as a weak limit of a sequence of probability measures μ_s on $\overline{B^3}$ that consists of atoms on the orbit $\Gamma(0)$. We define $\varphi_{\mu_s}(x)$ for these μ_s similarly to the case of measures on S^2. Then, even if we take the arrangement of the weight on $\Gamma(0)$ into account (which is necessary when Γ is of convergence type), we have $\varphi_{\mu_s}(x) = O(e^{(\delta(\Gamma)-\epsilon)\rho(0,x)})$ as in the proof of Lemma 12. Hence their weak limit μ cannot have an atom on ξ. □

We conclude this section with a summary of facts on geometrically finite Kleinian groups [**S1**], [**S4**].

THEOREM 13. *Let Γ be a geometrically finite Kleinian group. Then Γ is of divergence type and the unique $\delta(\Gamma)$-dimensional, Γ-invariant probability measure (PS measure) exists. Moreover, if Γ has no parabolic element, then the PS measure is coincident with the Hausdorff measure $h_{\delta(\Gamma)} = \mathcal{H}_{\delta(\Gamma)}|_{\Lambda_c(\Gamma)}$ up to a multiplicative constant. Furthermore, if μ_s is an s-dimensional, Γ-invariant probability measure* *with support on the limit set for $s > \delta(\Gamma)$, then μ_s consists only of atoms on bounded parabolic fixed points.*

PROOF. We take a $\delta(\Gamma)$-dimensional, Γ-invariant probability measure μ as in Theorem 11. Since μ has no atom on bounded parabolic fixed points, the total mass lies on $\Lambda_c(\Gamma)$. Hence, by Proposition 5, the $\delta(\Gamma)$-dimensional Poincaré series diverges. Also the uniqueness follows. In addition, if Γ has no parabolic elements, then $\mathcal{H}_{\delta(\Gamma)}(\Lambda_c(\Gamma)) > 0$ by the remark concerning the implication (3) $\Rightarrow$ (2). Hence the uniqueness implies that $h_{\delta(\Gamma)}$ coincides with μ modulo normalization. Next, by Proposition 5 again, if $s > \delta(\Gamma)$, then $\Lambda_c(\Gamma)$ is a null set for μ_s. Hence it has total mass on the bounded parabolic fixed points. Since they are countable, μ_s consists only of atoms. □

Using the fact that dim $\Lambda(\Gamma) =$ dim $\Lambda_c(\Gamma)$ for a geometrically finite Kleinian group Γ, we obtain the following corollary [**S4**], [**Tu1**].

COROLLARY 14. *The limit set $\Lambda(\Gamma)$ of a geometrically finite Kleinian group Γ has Hausdorff dimension 2 if and only if $\Lambda(\Gamma) = S^2$.*

PROOF. If $\Lambda(\Gamma) \neq S^2$ then the 2-dimensional Poincaré series converges. This can be seen from the fact that $\sum_{\gamma\in\Gamma} |\gamma'(x)|^2 < \infty$ for $x \in \Omega(\Gamma)$. On the other hand, Γ is of divergence type by Theorem 13. Therefore $\delta(\Gamma) < 2$, which implies that the Hausdorff dimension of the limit set is less than 2. □

5. Geometric indices

The Hausdorff dimension dim $\Lambda_c(\Gamma)$ of the conical limit set can be estimated by certain geometrical indices of the hyperbolic manifold N_Γ. More precisely, the Hausdorff dimension is related to them via the relationship between the critical exponent and the bottom of the spectrum.

DEFINITION (Bottom of spectrum). For a complete Riemannian manifold N in general, we define $\lambda_0(N)$ as the infimum of the *Rayleigh quotient*

$$\lambda_0(N) = \inf \left\{ \frac{\int_N |\nabla f|^2}{\int_N |f|^2} \;\middle|\; f \in C_0^\infty(N) \right\},$$

and call it the bottom of the spectrum. In particular, we denote the bottom of the spectrum of a hyperbolic manifold N_Γ by $\lambda_0(\Gamma)$.

The quantity $\lambda_0(N)$ is equal to the supremum (actually maximum) of the set of eigenvalues for smooth positive eigenfunctions with respect to the Laplace–Beltrami operator Δ on N. On the other hand, Δ is uniquely extendable to a semi-positive definite, self-adjoint operator on the Hilbert space of the square integrable functions on N, and $\lambda_0(N)$ is also equal to the infimum of the set of eigenvalues with respect to this operator. Geometrically, $\lambda_0(N)$ is related to the following constants.

DEFINITION (Isoperimetric constant, volume growth). For a Kleinian group Γ, we set

$$h(\Gamma) = \inf_W \frac{\mathrm{Area}\,(\partial W)}{\mathrm{Vol}\,(W)},$$

where the infimum is taken over all relatively compact subregions W of N_Γ, which we call the isoperimetric constant of N_Γ. Also we set

$$\kappa(\Gamma) = \limsup_{r\to\infty} \frac{1}{r} \log \mathrm{Vol}\,(B(p,r)),$$

where $B(p,r)$ is the subregion of N_Γ within a distance r of p.

Cheeger, Buser, and Brooks proved the following results respectively:

PROPOSITION 15.
[Ch]: $\lambda_0(\Gamma) \geq \frac{1}{4}h(\Gamma)^2$.
[Bu]: *There exists a universal constant C such that $\lambda_0(\Gamma) \leq Ch(\Gamma)$.*
[Br]: $\lambda_0(\Gamma) \leq \frac{1}{4}\kappa(\Gamma)^2$.

For a topologically tame hyperbolic manifold N_Γ, we can estimate the isoperimetric constant by considering the ratio of the volume of the convex core C_Γ to the area of its boundary; a geometrically infinite end (i.e., an end contained in C_Γ) is parabolic whereas a geometrically finite end (i.e., an end facing on $\Omega(\Gamma)$) is hyperbolic. The area of the boundary ∂C_Γ is bounded by a constant depending on the topological type of N_Γ (more concretely, depending on the minimal number of generators of Γ). Therefore, by estimating as Buser did, the following theorem due to Canary [**C1**] is obtained.

THEOREM 16. *A topologically tame hyperbolic manifold N_Γ satisfies*

$$\lambda_0(\Gamma) \leq \frac{K}{\mathrm{Vol}\,(C_\Gamma)},$$

where K is a constant depending only on the topological type of N_Γ. In particular, if N_Γ is topologically finite but not geometrically finite, then $\lambda_0(\Gamma) = 0$.

Hereafter in this section, we explain the following Elstrodt–Patterson–Sullivan theorem, which proves the relationship between the critical exponent and the bottom of the spectrum.

THEOREM 17. *Any Kleinian group Γ satisfies*

$$\lambda_0(\Gamma) = \begin{cases} 1 & (\delta(\Gamma) \leq 1), \\ \delta(\Gamma)(2-\delta(\Gamma)) & (\delta(\Gamma) > 1). \end{cases}$$

A proof of this theorem for Kleinian groups was given by Sullivan [**S5**]. Here we introduce an easier proof by assuming the Marden conjecture to be true. In particular, our proof is valid for Fuchsian groups.

For an s-dimensional, Γ-invariant probability measure μ, we consider a Γ-invariant function $\varphi_\mu(x) = \int_{S^2} P(x,\zeta)^s d\mu(\zeta)$. Since the Poisson kernel satisfies $\Delta P(x,\zeta)^s = s(2-s)P(x,\zeta)^s$, the projection of $\varphi_\mu(x)$ to N_Γ defines a positive eigenfunction for an eigenvalue $s(2-s)$ with respect to Δ. In particular, taking a $\delta(\Gamma)$-dimensional, Γ-invariant probability measure as μ, we obtain the following:

PROPOSITION 18. *There exists a positive eigenfunction on N_Γ for the eigenvalue $\delta(\Gamma)(2-\delta(\Gamma))$ with respect to Δ. Hence $\lambda_0(\Gamma) \geq \delta(\Gamma)(2-\delta(\Gamma))$.*

Next, we prove that Theorem 17 is valid for particular Kleinian groups by showing that the eigenfunctions are square integrable in those cases (see [**S4**], [**Pt3**]).

LEMMA 19. *If a geometrically finite Kleinian group Γ satisfies $\delta(\Gamma) > 1$, then the function φ_μ induced by the PS measure μ for Γ is square integrable on N_Γ.*

PROOF. The closed ϵ-neighborhood of the convex core C_Γ is denoted by C_Γ^*. The complement of the cusp neighborhoods in C_Γ^* is compact since Γ is geometrically finite. Suppose that Γ has a bounded parabolic fixed point ξ of rank k.

By Lemma 12, if x converges to ξ conically, then $\varphi_\mu(x) = O(e^{(k-\delta(\Gamma))\rho(0,x)})$. By Lemma 30, which will be proved later, $\epsilon = 2\delta(\Gamma) - k$ is positive, and thus $\varphi_\mu(x) = O(e^{(k-\epsilon)\rho(0,x)/2})$. From this we can see that φ_μ is square integrable in the cusp neighborhoods.

Next we will see that $\varphi_\mu(x)$ is also square integrable on $W = N_\Gamma - C^*_\Gamma$. For any point p in W, we take the nearest point p' on $\Sigma = \partial C^*_\Gamma$ to p and define a map $p \mapsto (p', t)$, where $t = \rho(p, p')$. This map induces a homeomorphism $h : W \to W' = \Sigma \times (0, \infty)$. Moreover, if we provide a metric $e^{2t} ds^2_\Sigma + dt^2$ for W' (where ds^2_Σ is the hyperbolic metric on Σ), then h is quasi-isometric (see [**C2**]). Hence we may use this metric instead in order to see that the integral is finite. Since $\varphi_\mu = O(e^{-t\delta(\Gamma)})$ as $t \to \infty$ by Lemma 12, we can reduce the problem to finding $\int_0^\infty e^{2t-2t\delta(\Gamma)} dt$. Then, by the assumption $\delta(\Gamma) > 1$, this is finite. □

Next we deal with a geometrically finite Kleinian group Γ with $\delta(\Gamma) \leq 1$. First of all, it is known that $\lambda_0(\{1\}) = 1$ for the trivial group. On the other hand, as we have seen above, if $\delta(\Gamma) \downarrow 1$, then $\lambda_0(\Gamma) \uparrow 1$. Hence it is natural to guess by interpolation that $\lambda_0(\Gamma)$ is identically equal to 1 whenever $0 < \delta(\Gamma) \leq 1$. Indeed, we will prove this assertion as follows. In general, if $\Gamma \subset G$, then $\lambda_0(\Gamma) \geq \lambda_0(G)$, because any positive eigenfunction on N_G with respect to Δ can be lifted to an eigenfunction on N_Γ for the same eigenvalue. Hence it suffices to prove that, for any $\epsilon > 0$, there exists a geometrically finite Kleinian group G that contains Γ and that satisfies $1 < \delta(G) < 1 + \epsilon$. To this end, relying on the following result due to Patterson [**Pt2**], we take the Klein combination G of Γ and a geometrically finite Kleinian group Γ' with $\delta(\Gamma') = 1$ (say a Fuchsian group Γ'). By Lemma 30, we can see that the critical exponent of G is greater than 1.

PROPOSITION 20. *For Kleinian groups Γ and Γ' with nonempty regions of discontinuity, there exists a divergent sequence g_n of Möbius transformations such that Γ and $g_n \Gamma' g_n^{-1}$ generate a Kleinian group $G_n = \Gamma * g_n \Gamma' g_n^{-1}$ by the Klein combination and that*

$$\lim_{n\to\infty} \delta(G_n) = \max\{\delta(\Gamma), \delta(\Gamma')\}.$$

So far we have proved that Theorem 17 is valid for a geometrically finite Kleinian group. In addition, Theorem 16 and Proposition 18 assert that, even if it is not geometrically finite, $\lambda_0(\Gamma) = 0$ and $\delta(\Gamma) = 2$ for a topologically tame hyperbolic manifold, and thus Theorem 17 is satisfied also in this case. Therefore, if we assume the Marden conjecture, Theorem 17 is valid for any finitely generated Kleinian group. Finally, to extend the result to the case of infinitely generated groups, we use the following lemma. Here, for a sequence of Kleinian groups Γ_n, we set

$$\mathrm{Env}\,\{\Gamma_n\} = \{\gamma \in \mathrm{PSL}_2(\mathbf{C}) \mid \gamma = \lim_{n\to\infty} \gamma_n \ (\gamma_n \in \Gamma_n)\}.$$

LEMMA 21. *Let Γ be a Kleinian subgroup of $\mathrm{Env}\,\{\Gamma_n\}$ for a sequence of Kleinian groups Γ_n. Then $\delta(\Gamma) \leq \liminf \delta(\Gamma_n)$. In particular, if a Kleinian group Γ is represented as $\Gamma = \bigcup_n \Gamma_n$ where $\Gamma_1 \subset \Gamma_2 \subset \cdots$, then $\delta(\Gamma) = \lim \delta(\Gamma_n)$ and $\lambda_0(\Gamma) = \lim \lambda_0(\Gamma_n)$.*

PROOF. Let μ_n be a $\delta(\Gamma_n)$-dimensional, Γ_n-invariant probability measure. Taking a subsequence, we may assume that the $\delta(\Gamma_n)$ converge to a dimension d, and taking a further subsequence, that the μ_n converge weakly to a measure μ on S^2.

Then μ is a d-dimensional, Γ-invariant probability measure. Thus the first assertion follows from $d \geq \alpha(\Gamma) = \delta(\Gamma)$. Under the assumption of the second assertion, we have $\delta(\Gamma) \geq \delta(\Gamma_n)$, and thus $\delta(\Gamma) = \lim \delta(\Gamma_n)$. Concerning $\lambda_0(\Gamma)$, for any $\epsilon > 0$ we take a function $f \in C_0^\infty(N_\Gamma)$ such that the Rayleigh quotient of f is less than $\lambda_0(\Gamma) + \epsilon$. Since the compact support of f is approximated by domains of N_{Γ_n}, we have a function $f_n \in C_0^\infty(N_{\Gamma_n})$ whose Rayleigh quotient is less than $\lambda_0(\Gamma) + 2\epsilon$ for any sufficiently large n. Hence $\lambda_0(\Gamma_n) \leq \lambda_0(\Gamma) + 2\epsilon$. Combining this inequality with $\lambda_0(\Gamma) \leq \lambda_0(\Gamma_n)$, we obtain the assertion. □

6. Geometric convergence and continuity of the Hausdorff dimension

We consider the change of the Hausdorff dimension of the limit set under deformation of a Kleinian group. It is seen from the distortion theorem on quasiconformal maps (see [**As**]) that the Hausdorff dimension varies continuously (with respect to the Teichmüller distance) under quasiconformal deformation. Moreover, it is a real analytic function on the quasiconformal deformation space for certain kinds of Kleinian groups [**Ru**], [**AR**]. Hence the problem lies in the continuity of the Hausdorff dimension on the boundary of the quasiconformal deformation spaces, and more generally for convergent sequences of Kleinian groups. To formulate this problem exactly, we begin by defining a couple of concepts on convergence of Kleinian groups.

DEFINITION (Algebraic convergence). Let Γ_0 be a group in general and let $\theta_n : \Gamma_0 \to \Gamma_n$ be a $\mathrm{PSL}_2(\mathbf{C})$-representation of Γ_0 onto Γ_n. We say that a sequence θ_n converges to a $\mathrm{PSL}_2(\mathbf{C})$-representation $\theta : \Gamma_0 \to \Gamma$ of Γ_0 onto Γ if the $\theta_n(\gamma)$ converge to $\theta(\gamma)$ for any $\gamma \in \Gamma_0$.

Hereafter, we always assume that Γ_0 is *finitely generated* when we consider algebraic convergence. Under this assumption, if the Γ_n are Kleinian, then their algebraic limit Γ is also a Kleinian group.

Although algebraic convergence preserves the algebraic structure of Kleinian groups, it does not necessarily imply convergence of the geometric structure of the corresponding hyperbolic manifolds. As we have seen in Section 5, the Hausdorff dimension of the limit set reflects the geometric nature of a hyperbolic manifold. Hence we have to take the following geometric convergence of a sequence of Kleinian groups into account in order to investigate the continuity of the Hausdorff dimension.

DEFINITION (Geometric convergence). We say that a sequence of subgroups Γ_n of $\mathrm{PSL}_2(\mathbf{C})$ converges geometrically to G if the Γ_n converge to G as closed subsets of $\mathrm{PSL}_2(\mathbf{C})$ in the *Hausdorff topology*; namely the following two conditions are satisfied:

- Any $g \in G$ can be written as $g = \lim_{n\to\infty} \gamma_n$ ($\gamma_n \in \Gamma_n$).
- Any element written as $g = \lim_{i\to\infty} \gamma_{n_i}$ ($\gamma_{n_i} \in \Gamma_{n_i}$) belongs to G.

The image of the origin $0 \in B^3$ under the projection $B^3 \to N_\Gamma = B^3/\Gamma$ is denoted by o_Γ. The condition that Kleinian groups Γ_n converge geometrically to G is equivalent to saying that the sequence of corresponding hyperbolic manifolds with the base point $(N_{\Gamma_n}, o_{\Gamma_n})$ converges to (N_G, o_G) *in the sense of Gromov*. That is, as n grows, larger neighborhoods of the base points in N_{Γ_n} and in N_G are mapped onto each other by diffeomorphisms that are closer to an isometry.

In general, a sequence of hyperbolic manifolds with base points has a convergent subsequence in the sense of Gromov if the injectivity radii at the base points are uniformly bounded from below. Hence, under certain circumstances that imply this condition, we can choose a geometrically convergent subsequence from a sequence of Kleinian groups. The assumption that the sequence is already known to be algebraically convergent produces one such situation.

PROPOSITION 22. *If representations $\theta_n : \Gamma_0 \to \Gamma_n$ onto Kleinian groups Γ_n converge algebraically to $\theta : \Gamma_0 \to \Gamma$, then a subsequence of Γ_n converges geometrically to a Kleinian group G that contains Γ.*

Bishop and Jones [**BJ**] proved lower semi-continuity of the Hausdorff dimension of the limit sets of finitely generated Kleinian groups under algebraic convergence.

THEOREM 23. *If representations $\theta_n : \Gamma_0 \to \Gamma_n$ onto Kleinian groups Γ_n converge algebraically to $\theta : \Gamma_0 \to \Gamma$, then*

$$\liminf_{n\to\infty} \dim \Lambda(\Gamma_n) \geq \dim \Lambda(\Gamma).$$

By Lemma 21, we can see that the Hausdorff dimension of the conical limit set has lower semi-continuity under either algebraic or geometric convergence. Hence the assertion of Theorem 23 immediately follows if the algebraic limit Γ is geometrically finite. From now on, in the case where Γ is not geometrically finite, we will prove a stronger assertion, the continuity of the Hausdorff dimension. The essential step for it is the following result, due to Bishop and Jones [**BJ**].

THEOREM 24. *If a finitely generated Kleinian group Γ is not geometrically finite, then* $\dim \Lambda(\Gamma) = 2$.

SKETCH OF PROOF. For the sake of simplicity, we explain this theorem in the case where the boundary ∂C_Γ of the convex core is totally geodesic. We take the double of C_Γ with respect to ∂C_Γ and denote it by N. Then N is represented as $N = N_G$ by a Kleinian group G. If $\lambda_0(\Gamma) = 0$, then $\dim \Lambda_c(\Gamma) = 2$ by Proposition 18. Thus we have only to consider the case $\lambda_0(\Gamma) > 0$. By the Ahlfors finiteness theorem, the hyperbolic area of ∂C_Γ is finite. By assumption, C_Γ has an infinite end. From these two facts, we can see that N_G satisfies $\lambda_0(G) > 0$, too. (We are convinced of this fact more easily if we consider the isoperimetric constant, for example.) This implies, in particular, that the harmonic measure of the ideal boundary of N_G is positive, and thus so is the harmonic measure of the ideal boundary of C_Γ, which is half of N_G divided along ∂C_Γ. This is equivalent to saying that the 2-dimensional measure of the limit set $\Lambda(\Gamma)$ is positive. In particular, $\dim \Lambda(\Gamma) = 2$. □

Note that if in addition N_Γ is known to be topologically tame, Theorem 16 already implied a stronger conclusion, $\dim \Lambda_c(\Gamma) = 2$, than Theorem 24.

REMARK. For a geometrically finite Kleinian group Γ, we have obtained the condition for $\dim \Lambda(\Gamma) = 2$ (Corollary 14). Then, with the aid of the theorem above, we can extend it to all finitely generated Kleinian groups. However, for infinitely generated Kleinian groups, such a condition has not yet been completely investigated. As a necessary condition, we may think of unboundedness of the distance of points in the convex core from its boundary [**Mt2**]. As we can see from the proof of Theorem 24, a sufficient condition is, say, analytic finiteness of $\Omega(\Gamma)/\Gamma$, which is a consequence of the Ahlfors finiteness theorem [**BJ**].

Therefore, by the following lemma, we complete a proof of Theorem 23.

LEMMA 25. *If representations $\theta_n : \Gamma_0 \to \Gamma_n$ onto Kleinian groups Γ_n converge algebraically to $\theta : \Gamma_0 \to \Gamma$ and if Γ is not geometrically finite, then*

$$\lim_{n\to\infty} \dim \Lambda(\Gamma_n) = \dim \Lambda(\Gamma) \ (= 2).$$

PROOF. Passing to a subsequence, we may assume that the Γ_n are all geometrically finite or all geometrically infinite. In the latter case, the assertion is clear by Theorem 24. Hence we only consider the case where the Γ_n are all geometrically finite. Passing to a subsequence again, we may assume that the $\dim \Lambda(\Gamma_n)$ converge and, by Proposition 22, that the Γ_n converge geometrically to a Kleinian group G that contains Γ. If $\lim \dim \Lambda(\Gamma_n) < 2$, then, by Theorem 16, the volumes of the convex cores C_{Γ_n} are uniformly bounded from above. In this case, the volume of C_G is finite (see [**Ta**]), and thus G is geometrically finite. By lower semi-continuity, $\dim \Lambda(G) = \dim \Lambda_c(G) < 2$. However, this contradicts $\dim \Lambda(\Gamma) = 2$. Hence $\lim \dim \Lambda(\Gamma_n) = 2$, which implies the assertion. □

REMARK. On the other hand, when a sequence of Kleinian groups Γ_n converges geometrically to G, we have a similar assertion; if G is finitely generated but not geometrically finite, then $\lim \dim \Lambda(\Gamma_n) = \dim \Lambda(G)$. Indeed, if the geometric limit is finitely generated in general, then there exist homomorphisms (not necessarily surjective) $\psi_n : G \to \Gamma_n$ for all sufficiently large n that converge algebraically to the identity isomorphism $\mathrm{id} : G \to G$ [**JM**]. Then applying Lemma 25, we obtain this assertion.

From these observations, we can see that the case where the limit is geometrically finite is essential as long as we consider the problem on the convergence of the Hausdorff dimension. Thus we restrict ourselves to this case hereafter.

First of all, we exhibit a simple example where the geometric convergence does not necessarily imply the convergence of the Hausdorff dimension of the limit set.

EXAMPLE. We take a sequence of finite index subgroups Γ_n of a Kleinian group Γ_0 such that $\Gamma_0 \supset \Gamma_1 \supset \Gamma_2 \cdots$ converge geometrically to $\{1\}$. (For example, we can take such a sequence by letting the injectivity radii at the base point increase to ∞.) Then $\dim \Lambda(\Gamma_n) = \dim \Lambda(\Gamma_0) > 0$, but $\dim \Lambda(\{1\}) = 0$. The limit sets $\Lambda(\Gamma_n) = \Lambda(\Gamma_0)$ do not converge to $\Lambda(\{1\}) = \emptyset$ in the Hausdorff topology either.

In order to show the convergence of the Hausdorff dimension, we plan the following strategy: Let μ be a weak limit of the PS measures μ_n for geometrically finite Kleinian groups Γ_n. If we were able to determine that μ is the PS measure for the geometric limit G, then the critical exponents $\delta(\Gamma_n)$ would converge to $\delta(G)$, and thus $\dim \Lambda(\Gamma_n) \to \dim \Lambda(G)$ would be obtained. However, even though μ is a G-invariant probability measure, it may have a positive mass on $\Omega(G)$ in general. If we know at least that the $\Lambda(\Gamma_n)$ converge to $\Lambda(G)$ in the Hausdorff topology, we can see that μ has its support on $\Lambda(G)$, which is a necessary condition for μ to be the PS measure. Hence, in order to keep this line, we have to impose a condition such that at least the Hausdorff convergence of the limit set is guaranteed.

REMARK. However, even if the limit sets do not converge in the Hausdorff topology, the Hausdorff dimension may converge. For example, we can take the Klein combination $G_n = \Gamma * g_n \Gamma' g_n^{-1}$ so that it converges geometrically to Γ as in Proposition 20, but the $\Lambda(G_n)$ do not converge to $\Lambda(\Gamma)$ in the Hausdorff topology.

(Remark that the assumption of Corollary 7.34 in [**MT**] is insufficient.) If $\delta(\Gamma) < \delta(\Gamma')$, then the Hausdorff dimension does not converge (and the weak limit μ of the PS measures has a positive mass on $\Omega(\Gamma)$), whereas if $\delta(\Gamma) \geq \delta(\Gamma')$, then it converges.

When the Hausdorff dimension of the limit set converges and the geometric limit G is a geometrically finite Kleinian group without a parabolic element, we can see the convergence of the Hausdorff dimension, because the only G-invariant probability measure with support on the limit set is the PS measure. However, when G contains a parabolic element, the Hausdorff convergence of the limit sets does not necessarily imply the convergence of the Hausdorff dimension. In more detail, the weak limit μ of the PS measures may consist of atoms on parabolic fixed points of G, and in this case $\delta(G)$ must be less than the dimension of μ. For example, if we explain this phenomenon in terms of hyperbolic manifolds, it occurs in the case where a Gromov convergent sequence of hyperbolic manifolds collapses at a cusp so that the "main part" disappears in the limit and a mass of the measure that was in the main part is concentrated on the parabolic fixed points that correspond to the new cusp.

Therefore, to make the geometric structure of hyperbolic manifolds converge under geometric convergence and to make the structure of the fundamental group be preserved (i.e., to make the manifold not collapse at the limit) under algebraic convergence, we define the following convergence, which includes both properties.

DEFINITION (Strong convergence). We say that a sequence of Kleinian groups Γ_n converges strongly to G if the Γ_n converge geometrically to G and, for all sufficiently large n, there exist surjective homomorphisms $\psi_n : G \to \Gamma_n$ that converge algebraically to $\mathrm{id} : G \to G$.

REMARK. Usually we define strong convergence by requiring that an algebraically convergent sequence of faithful discrete representations is geometrically convergent at the same time. Clearly our definition above, which is due to McMullen [**Mc**], is wider than the usual one. Moreover, even when the algebraic limit and the geometric limit differ, if the geometric limit is finitely generated, then it is strongly convergent in our sense. Indeed, although we mentioned that there exist homomorphisms, not necessarily surjective, of a finitely generated geometric limit that converge algebraically to the identity isomorphism, they are actually surjective if we know in addition that the original sequence is algebraically convergent [**JM**].

If the limit G of a strongly convergent sequence is geometrically finite, then the Γ_n for all sufficiently large n are also geometrically finite and $\Lambda(\Gamma_n)$ converge to $\Lambda(G)$ in the Hausdorff topology [**JM**]. However, even in this case, there still exists an example, raised by McMullen [**Mc**], where the Hausdorff dimension does not converge.

Concerning convergence of the bottoms of the spectrum of hyperbolic manifolds under strong convergence, Canary and Taylor [**CT2**] showed the following result by a proof of geometric flavor independently of McMullen. Also, Fan and Jorgenson [**FJ**] proved convergence of the heat kernels of hyperbolic manifolds and obtained a similar result.

THEOREM 26. *If a sequence of Kleinian groups Γ_n converges strongly to G and if N_G is topologically tame, then* $\lim_{n\to\infty} \lambda_0(\Gamma_n) = \lambda_0(G)$.

The reason why the convergence of the bottom of the spectrum does not necessarily imply the convergence of the Hausdorff dimension lies in their relationship (Theorem 17); as long as the critical exponent is not greater than 1, the bottom of the spectrum is identically 1. In fact, McMullen's counterexample is a phenomenon that occurs near the dimension 1. He proved the following theorem concerning the convergence of the Hausdorff dimension under strong convergence by an analytic method, and obtained the theorem above contrarily by interpreting the Hausdorff dimension into the bottom of the spectrum [**Mc**].

THEOREM 27. *If a sequence of Kleinian groups* Γ_n *converges strongly to a finitely generated Kleinian group* G *that satisfies* $\dim \Lambda(G) \geq 1$, *then*

$$\lim_{n\to\infty} \dim \Lambda(\Gamma_n) = \dim \Lambda(G).$$

7. Hausdorff dimension 1

Hausdorff dimension 1 stands at a special position among the dimensions of the limit sets of Kleinian groups. Actually, we have experienced certain arguments in which the Hausdorff dimension and the critical exponent have a different feature on different sides of 1. The Hausdorff dimension of the limit set of a Fuchsian group is clearly 1, and conversely Bowen [**Bo**] proved that only a Fuchsian group is a quasi-Fuchsian group whose limit set has Hausdorff dimension 1. Generalizing this result, we completely determine finitely generated Kleinian groups whose limit sets have Hausdorff dimension not greater than 1 [**CT1**], [**BJ**].

DEFINITION (Quasi-Fuchsian group). A geometrically finite Kleinian group Γ such that $\Omega(\Gamma)$ consists of two Γ-invariant Jordan domains is called a quasi-Fuchsian group. In particular, if the two Jordan domains are round disks, it is called a Fuchsian group.

THEOREM 28. *If a quasi-Fuchsian group has the limit set of Hausdorff dimension 1 then it is a Fuchsian group.*

PROOF. Let Γ be a quasi-Fuchsian group whose limit set has Hausdorff dimension 1. By Lemma 6, the 1-dimensional Hausdorff measure of $\Lambda(\Gamma)$ is finite—in other words, $\Lambda(\Gamma)$ is rectifiable. Then the following lemma implies that Γ is a Fuchsian group. □

LEMMA 29. *If the limit set of a quasi-Fuchsian group* Γ *is rectifiable, then* Γ *is a Fuchsian group.*

PROOF. Let Ω_1 and Ω_2 be the two invariant components of $\Omega(\Gamma)$, and consider Riemann mappings $f_1 : \Delta \to \Omega_1$ and $f_2 : \Delta^* \to \Omega_2$, where Δ is the unit disk and $\Delta^* = \hat{\mathbf{C}} - \overline{\Delta}$. Then f_1^{-1} and f_2^{-1} induce conjugation of Γ onto Fuchsian groups Γ_1 and Γ_2, respectively. Since f_1 and f_2 extend to the boundary homeomorphically, a homeomorphism $(f_2)^{-1} \circ f_1$ of $\partial\Delta$ is defined. By the assumption that $\Lambda(\Gamma)$ is rectifiable, the 1-dimensional Hausdorff measure and the harmonic measure on $\Lambda(\Gamma)$ are mutually absolutely continuous with respect to each other. Therefore $(f_2)^{-1} \circ f_1$ is absolutely continuous with respect to the 1-dimensional measure on the unit circle that induces conjugation between Fuchsian groups Γ_1 and Γ_2. Here we apply the rigidity of Fuchsian groups: an automorphism of the unit circle that is compatible with a Fuchsian group is either (totally) singular with respect to the 1-dimensional measure or the restriction of a Möbius transformation. Then

$(f_2)^{-1} \circ f_1$ is the restriction of a Möbius transformation, from which we see that Γ should be a Fuchsian group. □

REMARK. A quasi-Fuchsian group is alternatively defined as a quasiconformal deformation of a Fuchsian group. For a quasiconformal deformation of a Fuchsian group in general (which is not necessarily finitely generated), we may consider the statement of Lemma 29, but this is not true in general. The reason why not is that the rigidity of Fuchsian groups, which was used in the above proof, is valid only for a Fuchsian group whose 1-dimensional Poincaré series diverges (see [**Mt1**]). Astala and Zinsmeister showed the necessity of this condition, as well as an example where a nontrivial quasiconformal deformation of a Fuchsian group has a rectifiable limit set [**AZ**].

The following lemma is useful not only for the extension of Theorem 28 but also at several places where we have utilized it already.

LEMMA 30. *If a Kleinian group* Γ *contains a Kleinian subgroup* H *of divergence type and if* $\Lambda(\Gamma)$ *properly contains* $\Lambda(H)$, *then* $\delta(\Gamma) > \delta(H)$. *In particular, if* Γ *contains a parabolic abelian subgroup of rank* $k = 1$ *or* 2, *then* $\delta(\Gamma) > k/2$.

PROOF. Let μ be a $\delta(\Gamma)$-dimensional, Γ-invariant probability measure. We can take a disk $\Delta = \Delta(x, r) \subset \Omega(H)$ such that $\mu(\Delta) > 0$ and $\Delta \cap h(\Delta) = \emptyset$ for any nontrivial $h \in H$. Since

$$1 > \sum_{h \in H} \mu(h(\Delta)) \approx \mu(\Delta) \cdot \sum_{h \in H} |h'(x)|^{\delta(\Gamma)},$$

the $\delta(\Gamma)$-dimensional Poincaré series for H converges. Since H is of divergence type, this implies that $\delta(\Gamma) > \delta(H)$.

For a parabolic abelian subgroup J of rank k, the Poincaré series $\sum_{j \in J} |j'(z)|^s$ $(z \in \Omega(J))$ converges for $s > k/2$ and diverges for $s \leq k/2$ (see Lemma 12). Hence we obtain the latter assertion. □

THEOREM 31. *For a finitely generated Kleinian group* Γ, *if* $\dim \Lambda(\Gamma) \leq 1$, *then* $\Lambda(\Gamma)$ *is either a circle or a totally disconnected set.*

PROOF. In general, a finitely generated Kleinian group Γ has either a quasi-Fuchsian subgroup, a totally degenerate subgroup or a totally disconnected limit set. However, it cannot have a totally degenerate subgroup because it has the limit set of Hausdorff dimension 2 by Theorem 16 or 24. If Γ has a quasi-Fuchsian subgroup H, it follows from Theorem 28 that H is Fuchsian. Then Lemma 30 implies that $\Lambda(\Gamma) = \Lambda(H)$. □

COROLLARY 32. *If* $\dim \Lambda(\Gamma) < 1$ *for a finitely generated Kleinian group* Γ, *then the hyperbolic manifold* N_Γ *is homeomorphic to the interior of a handlebody.*

PROOF. By Theorem 31, $\Lambda(\Gamma)$ is a totally disconnected set. If Γ contained a parabolic abelian subgroup of rank 2, then we would have $\dim \Lambda(\Gamma) > 1$ by Lemma 30. Thus Γ must be a Schottky-like group which may contain a parabolic cyclic subgroup. This is equivalent to the conclusion of the statement. □

From this result, we see that the Hausdorff dimension determines the topology of a hyperbolic manifold in some case. Conversely, we define a topological invariant of a 3-manifold from the Hausdorff dimension.

DEFINITION. For a compact 3-manifold M (with boundary) that admits a hyperbolic structure, we define $D(M)$ as the infimum of dim $\Lambda(\Gamma)$ taken over all Kleinian groups Γ such that the hyperbolic manifold N_Γ is homeomorphic to the interior of M.

By Corollary 32, if $D(M) = 0$, then M is a handlebody. Conversely, if M is a handlebody, then $D(M) = 0$ by Proposition 20. Hence, the next possible value of $D(M)$ is 1. Though it had been known that if M is an I-bundle (trivial or twisted bundle) over a surface, then $D(M) = 1$, Canary, Minsky and Taylor [**CMT**] completely determined all other 3-manifolds M with $D(M) = 1$. Intuitively speaking, if M is a union of parts of the product structure, we can take the following sequence, which shows $D(M) = 1$ by reasoning similar to Proposition 20: the Hausdorff dimension of a quasi-Fuchsian group that corresponds to each part of M converges to 1, but only one of them remains and the others disappear, so that M converges geometrically to a Fuchsian product structure. Conversely, they proved $D(M) > 1$ for the other 3-manifolds M. Here we show a proof of this fact only in the simplest case.

PROPOSITION 33. *If M is acylindrical and not an I-bundle, then $D(M) > 1$.*

PROOF. Suppose that there exists a sequence of Kleinian groups Γ_n such that N_{Γ_n} is homeomorphic to the interior of M and dim $\Lambda(\Gamma_n) \to 1$. By compactness of the set of discrete faithful representations of the fundamental group of an acylindrical M [**Th**], we may assume that the $\theta_n : \pi_1(M) \to \Gamma_n$ converge algebraically to a discrete faithful representation $\theta : \pi_1(M) \to \Gamma$. Then dim $\Lambda(\Gamma) \leq 1$ by Theorem 23; however, this contradicts Theorem 31. □

References

[AR] J. Anderson and A. Rocha, *Analyticity of Hausdorff dimension of limit sets of Kleinian groups,* Ann. Acad. Sci. Fenn. **22** (1997), 349–364.

[Ah] L. Ahlfors, *Möbius transformations in several dimensions,* Lecture Notes at University of Minnesota, 1981.

[As] K. Astala, *Area distortion of quasiconformal mappings,* Acta Math. **173** (1994), 37–60.

[AZ] K. Astala and M. Zinsmeister, *Abelian coverings, Poincaré exponent of convergence, and holomorphic deformations,* Ann. Acad. Sci. Fenn. **20** (1995), 81–86.

[BJ] C. Bishop and P. Jones, *Hausdorff dimension and Kleinian groups,* Acta Math. **179** (1997), 1–39.

[Bo] R. Bowen, *Hausdorff dimension of quasi-circles,* Inst. Hautes Études Sci. Publ. Math. **50** (1979), 11–26.

[Br] R. Brooks, *A relation between growth and the spectrum,* Math. Z. **178** (1981), 501–508.

[Bu] P. Buser *A note on the isoperimetric constant,* Ann. Sci. École Norm. Sup. **15** (1982), 213–230.

[C1] R. Canary, *On the Laplacian and the geometry of hyperbolic 3-manifolds,* J. Diff. Geom. **36** (1992), 349–367.

[C2] R. Canary, *Ends of hyperbolic 3-manifolds,* J. Amer. Math. Soc. **6** (1993), 1–35.

[CT1] R. Canary and E. Taylor, *Kleinian groups with small limit sets,* Duke Math. J. **73** (1994), 371–381.

[CT2] R. Canary and E. Taylor, *Hausdorff dimension and limits of Kleinian groups,* Geom. Funct. Anal. **9** (1999), 283–297.

[CMT] R. Canary, Y. Minsky, and E. Taylor, *Spectral theory, Hausdorff dimension and the topology of hyperbolic 3-manifolds,* J. Geom. Anal. **9** (1999), 17–40.

[Ch] J. Cheeger, *A lower bound for the smallest eigenvalue of the Laplacian,* in *Problems in Analysis,* 195–199, Princeton Univ. Press, 1970.

[Fa] K. Falconer, *The Geometry of Fractal Sets,* Cambridge Univ. Press, 1985.

[FJ] C. Fan and J. Jorgenson, *Small eigenvalues and Hausdorff dimension of sequences of hyperbolic 3-manifolds,* preprint.

[Fe] H. Federer, *Geometric Measure Theory,* Springer-Verlag, Berlin, 1969.

[JM] T. Jørgensen and A. Marden, *Algebraic and geometric convergence of Kleinian groups,* Math. Scand. **66** (1990), 47–72.

[Mg] W. Magnus, *Noneuclidean Tesselations and Their Groups,* Pure and Applied Mathematics, Vol. 61, Academic Press, New York-London, 1974.

[Mt1] K. Matsuzaki *Several characterizations of Fuchsian groups of divergence type,* in *Complex Analysis on Hyperbolic 3-manifolds,* RIMS Kokyuroku **882** (1994), 51–56. MR **96h:**30084

[Mt2] K. Matsuzaki *Hausdorff dimension of the limit sets of infinitely generated Kleinian groups,* Math. Proc. Camb. Phil. Soc. **128** (2000), 123–139.

[MT] K. Matsuzaki and M. Taniguchi, *Hyperbolic manifolds and Kleinian groups,* Oxford Univ. Press, 1998.

[Mc] C. McMullen, *Hausdorff dimension and conformal dynamics I: Strong convergence of Kleinian groups,* J. Diff. Geom. **51** (1999), 471–515.

[N] P. Nicholls, *The ergodic theory of discrete groups,* London Math. Soc. Lecture Note Series **143**, Cambridge Univ. Press, 1989.

[O] K. Ohshika, *An approach to Marden's conjecture for finitely generated Kleinian groups,* in *Analysis of Discrete Groups,* RIMS Kokyuroku **967** (1996), 100–101.

[Pt1] S. Patterson, *The limit set of a Fuchsian group,* Acta Math. **136** (1976), 241–273.

[Pt2] S. Patterson, *Further remarks on the exponent of convergence of Poincaré series,* Tohoku Math. J. **35** (1983), 357–373.

[Pt3] S. Patterson, *Lectures on measures on limit sets of Kleinian groups,* in *Analytical and Geometric Aspects of Hyperbolic Space,* 281–323. London Math. Soc. Lecture Note Series **111**, Cambridge Univ. Press, 1987.

[Po] H. Poincaré, *Papers on Fuchsian Functions,* Springer-Verlag, New York, 1985.

[Ro] C. Rogers, *Hausdorff Measures,* Cambridge Univ. Press, 1970.

[Ru] D. Ruelle, *Repellers for real analytic maps,* Ergodic Theory and Dynamical Systems **2** (1982), 99–107.

[S1] D. Sullivan, *The density at infinity of a discrete group of hyperbolic motions,* Inst. Hautes Études Sci. Publ. Math. **50** (1979), 172–202.

[S2] D. Sullivan, *On the ergodic theory at infinity of an arbitrary discrete group of hyperbolic motions,* Riemann Surfaces and Related Topics: Proc. 1978 Stony Brook Conf. (I. Kra and B. Maskit, eds.), Ann. of Math. Studies, vol. 97, Princeton Univ. Press, Princeton, NJ, 1981, pp. 465–496.

[S3] D. Sullivan, *Discrete conformal groups and measurable dynamics,* Bull. Amer. Math. Soc. **6** (1982), 57–73.

[S4] D. Sullivan, *Entropy, Hausdorff measures old and new, and limit sets of geometrically finite Kleinian groups,* Acta Math. **153** (1984), 259–277.

[S5] D. Sullivan, *Related aspects of positivity in Riemannian geometry,* J. Diff. Geom. **25** (1987), 327–351.

[Ta] E. Taylor, *Geometric finiteness and the convergence of Kleinian groups,* Comm. Anal. Geom. **5** (1997), 497-533.

[Th] W. Thurston, *Hyperbolic structures on 3-manifolds I: Deformation of acylindrical 3-manifolds,* Ann. of Math. **124** (1986), 203–246.

[Tu1] P. Tukia, *The Hausdorff dimension of the limit sets of geometrically finite Kleinian groups,* Acta Math. **152** (1984), 127–140.

[Tu2] P. Tukia, *The Poincaré series and conformal measure of conical and Myrberg limit points,* J. Analyse Math. **62** (1994), 241–259.

Department of Mathematics, Ochanomizu University, Otsuka, Bunkyo-ku, Tokyo 112-8610, Japan

E-mail address: `matsuzak@math.ocha.ac.jp`

Translated by KATSUHIKO MATSUZAKI

Amer. Math. Soc. Transl.
(2) Vol. **204**, 2001

Topological Methods in Stability Analysis of Travelling Waves

Shunsaku Nii

1. Introduction

Travelling wave solutions, which are solutions that translate spatially at constant velocity preserving their profile through time, have special importance from the physical viewpoint. These solutions arise in nerve axon equations, which are model equations for the propagation of a pulse in nerve axons; reaction-diffusion equations, which describe spatially inhomogeneous chemical reactions or spatial distribution of population density of species; and so forth. In the case of nerve equations, for example, these solutions are supposed to correspond to electric pulses that propagate along nerve axons. (See, e.g., Murray [**32**] for more detail.)

One problem arises, however, in the study of travelling waves in this context. It is difficult to observe unstable solutions, which are easily destroyed by small perturbations of their initial conditions, in the real phenomena described, because real systems are exposed to noise from outside the systems. Therefore, it is necessary to carry out a stability analysis of each wave, that is, to find out whether the solution returns to the original one with increasing time even if its initial condition is slightly perturbed.

One way to prove the stability of a given wave is to analyse the linearised eigenvalue problem. Evans [**6, 7, 8, 9**] formulated the stability problem of waves in nerve equations and defined an analytic function, called Evans' function after him, which vanishes at each eigenvalue and whose order of zero coincides with the multiplicity of the eigenvalue. Evans' function has been used in the study of several systems other than nerve equations, and has become one of the standard methods for analysing eigenvalue problems. (See Jones [**24**], Ikeda [**20**], Maginu [**29**], Yanagida [**48**], Yanagida-Maginu [**49**], Pego-Weinstein [**40**], and others.)

Although the proof of stability of pulse solutions in the FitzHugh-Nagumo equations in one of the above papers, Jones [**24**], was based on Evans' function, the method used was not calculation of zeros of Evans' function analytically. Instead, the approach was based on a topological aspect of Evans' function, more precisely, the fact that the number of eigenvalues in a given domain in the complex plane

This article originally appeared in Japanese in Sūgaku **51** (3) (1999), 241–256.
2000 *Mathematics Subject Classification.* Primary 35K57; Secondary 35B25, 35B35, 37L10, 92C20.

is equal to the winding number of the image of the boundary of the domain under Evans' function, because Evans' function is analytic. This argument laid the foundation of a general topological framework for dealing with eigenvalue problems, called the stability index, which is the main topic of this paper.

Despite Evans' function having been developed for the stability analysis of pulses in nerve equations, its topological reformulation, stability index, has a wide range of applications, not only to the singular perturbation problem or to bifurcation problems of general reaction-diffusion equations of bistable type, which it is naturally applied to, but also to spectrum analysis for linearisation along periodic travelling waves (Gardner [**12**]), the linearising eigenvalue problem of steady state solutions in finite intervals (Gardner-Jones [**18**]), and so forth. Moreover, a recent result (Gardner [**14**], [**15**]) shows that by making Evans' function multivalued, it can be extended to cases where an unstable Fisher type travelling wave appears in the construction of the wave by singular perturbation.

The purpose of this paper is to introduce the theory of the stability index, including recent developments.

REMARK 1. Because of limited space, we do not explain topological terminology in this article. Readers are referred to Milnor-Stasheff [**31**], Steenrod [**46**] and Tamura [**47**].

2. Travelling waves of reaction-diffusion equations and the stability index

In this section, the construction of Evans' function and the stability index is explained. For the sake of simplicity, the systems treated are restricted to reaction-diffusion equations.

2.1. Travelling waves and their asymptotic stability. Consider a system of reaction-diffusion equations in one space variable:

$$u_t = Bu_{xx} + f(u) \qquad (t > 0,\ -\infty < x < +\infty,\ u(t,x) \in \mathbb{R}^n), \tag{1}$$

where B is a positive diagonal matrix, and $f\colon \mathbb{R}^n \to \mathbb{R}^n$ is of class C^2 having zeros $u_\pm$. Moreover, the steady state solutions $u \equiv u_\pm$ of (1) are assumed to be linearly stable, that is, the following holds for the linearisation of (1) at $u \equiv u_\pm$:

$$p_t = L_\pm p := Bp_{xx} + Df(u_\pm)p.$$

HYPOTHESIS 1. Let $\sigma(L_\pm)$ be the spectrum of $L_\pm$. Then there exists a negative β such that $\sigma(L_\pm) \subset \{\lambda \mid \operatorname{Re}\lambda < \beta\}$.

Let $\xi = x - \theta t$ for some constant θ and rewrite (1) in (t, ξ) coordinates:

$$u_t = Bu_{\xi\xi} + \theta u_\xi + f(u). \tag{2}$$

Because this is the system (1) expressed in the moving frame with the velocity θ, a stationary solution $u(t,\xi) = u(\xi)$ of (2) corresponds to a solution of (1) that translates in the velocity θ preserving its profile. This is a travelling wave (Figure 1). In what follows, existence of a travelling wave with the following conditions is assumed.

HYPOTHESIS 2. There exist $C > 0$ and $a > 0$ with the properties

1. $|u(\xi) - u_-| < Ce^{a\xi} \quad (\xi < 0)$ and $|u(\xi) - u_+| < Ce^{-a\xi} \quad (\xi > 0)$,
2. $|u'(\xi)| < Ce^{-a|\xi|} \quad (-\infty < \xi < +\infty)$, where $' = \frac{d}{d\xi}$.

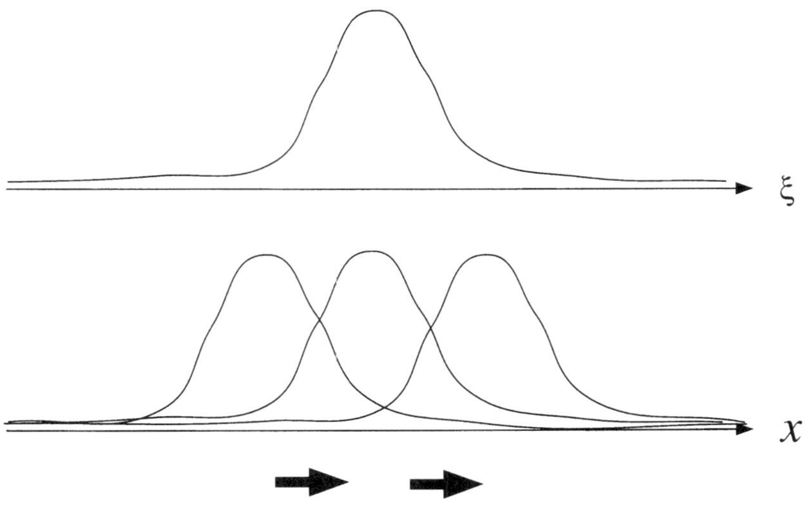

FIGURE 1

This is equivalent to the existence of a solution $(u(\xi), u'(\xi))$ of the system below with $\lim_{\xi\to-\infty}(u(\xi), u'(\xi)) = (u_-, 0)$ and $\lim_{\xi\to+\infty}(u(\xi), u'(\xi)) = (u_+, 0)$:

$$\begin{cases} u' = v, \\ v' = -B^{-1}[f(u) + \theta v], \end{cases} \qquad (' = \tfrac{d}{d\xi}). \tag{3}$$

This system is obtained by rewriting the second-order system

$$Bu_{\xi\xi} + \theta u_\xi + f(u) = 0 \tag{4}$$

into a first-order system.

Asymptotic stability of a travelling wave is defined as follows.

DEFINITION 1. A travelling wave $u(\xi)$ is asymptotically stable if there exists a neighbourhood N of $u(\xi)$ in $BU(\mathbb{R}, \mathbb{R}^n)$ such that any solution $\tilde{u}(t, \xi)$ of (2) with $\tilde{u}(t, \xi) \in N$ satisfies the following for some $k \in \mathbb{R}$:

$$\|\tilde{u}(t, \xi) - u(\xi + k)\|_\infty \to 0 \quad (t \to \infty), \tag{5}$$

where $BU(\mathbb{R}, \mathbb{R}^n)$ denotes the space of bounded and uniformly continuous functions from $\mathbb{R}$ to $\mathbb{R}^n$ supplied with the supremum norm.

REMARK 2. If $u(\xi)$ is a travelling wave, then so is its translation $u(\xi + k)$.

Namely, $u(\xi)$ is stable if it recovers to its spatial translation even if it is perturbed slightly.

The following gives a sufficient condition.

FACT 1 ([**4**], *etc.*). *Let* $p_t = Lp := Bp_{\xi\xi} + \theta p_\xi + Df(u(\xi))p$ *be the linearisation of* (2) *along* $u(\xi)$. *Then* $u(\xi)$ *is asymptotically stable if the spectrum* $\sigma(L)$ *of* L *satisfies the following:*

1. $\sigma(L) \setminus \{0\} \subset \{\lambda \mid \operatorname{Re}\lambda < \beta\}$, *where* $\beta < 0$ *is from hypothesis* 1;
2. 0 *is a simple eigenvalue of* L.

REMARK 3. 0 is an eigenvalue of L corresponding to the translational freedom of $u(\xi)$, and the corresponding eigenfunction is $u'(\xi)$.

This fact together with the following proposition means that the stability problem amounts to the existence problem of eigenvalues in some bounded domain.

PROPOSITION 1 ([**1**]). *There is a simple closed curve K with the following property: If K° is the domain enclosed by K, then $\sigma(L) \cap \{\lambda \mid \operatorname{Re}\lambda > \beta\} \subset K^\circ$, and $\sigma(L) \cap K^\circ$ consists of isolated eigenvalues with finite multiplicity.*

Therefore, the stability problem of a given travelling wave becomes the eigenvalue problem of L. One way to tackle this problem is to use Evans' function, which Evans [**9**] developed to prove instability of pulses in nerve equations.

2.2. Evans' function. The eigenvalue problem for a second-order ordinary differential operator L,

$$Lp = Bp_{\xi\xi} + \theta p_\xi + Df\,(u(\xi))\,p = \lambda p, \tag{6}$$

can be regarded as a second-order ordinary differential equation:

$$\begin{pmatrix} p \\ q \end{pmatrix}' = \begin{pmatrix} 0 & I \\ B^{-1}\,(\lambda I - Df\,(u(\xi))) & -\,\theta B^{-1} \end{pmatrix} \begin{pmatrix} p \\ q \end{pmatrix}, \tag{7}$$

where $'$ denotes differentiation with respect to ξ. This is simply written as

$$Y' = A(\xi;\lambda)Y.$$

Since the problem is considered in the space of bounded functions, λ's being an eigenvalue of L is equivalent to (7)'s having a nontrivial bounded solution as $\xi \to \pm\infty$.

$A(\xi;\lambda)$ satisfies

$$\lim_{\xi\to\pm\infty} A(\xi;\lambda) = A_\pm(\lambda) := \begin{pmatrix} 0 & I \\ B^{-1}\,(\lambda I - Df\,(u_\pm)) & -\,\theta B^{-1} \end{pmatrix}, \tag{8}$$

and under hypothesis 1, $A_\pm(\lambda)$ has n eigenvalues with positive real part and n eigenvalues with negative real part if $\operatorname{Re}\lambda > \beta$. This means that (7) has n independent solutions $Y_1(\xi;\lambda), \cdots, Y_n(\xi;\lambda)$ converging to 0 as $\xi \to -\infty$ and n independent solutions $Y_{n+1}(\xi;\lambda), \cdots, Y_{2n}(\xi;\lambda)$ converging to 0 as $\xi \to +\infty$. On the other hand, any solution except linear combinations of $Y_1(\xi;\lambda), \cdots, Y_n(\xi;\lambda)$ diverges as $\xi \to -\infty$, and any solution except linear combinations of $Y_{n+1}(\xi;\lambda), \cdots, Y_{2n}(\xi;\lambda)$ diverges as $\xi \to +\infty$. Therefore the following holds.

LEMMA 1. *λ is an eigenvalue of L if and only if*

$$Y_1(\xi;\lambda), \cdots, Y_n(\xi;\lambda), Y_{n+1}(\xi;\lambda), \cdots, Y_{2n}(\xi;\lambda)$$

are linearly dependent.

This is because, if $Y(\xi,\lambda)$ is an eigenfunction of L belonging to the eigenvalue λ, then $Y(\xi;\lambda)$ is a linear combination of $Y_1(\xi;\lambda), \cdots, Y_n(\xi;\lambda)$, since it is bounded as $\xi \to -\infty$; at the same time, it is also a linear combination of $Y_{n+1}(\xi;\lambda), \cdots, Y_{2n}(\xi;\lambda)$, since it is bounded as $\xi \to +\infty$.

Having seen this, we also see that the following definition of Evans' function is natural.

DEFINITION 2. The function

$$D(\lambda) := \det\left(Y_1(0;\lambda), \cdots, Y_n(0;\lambda), Y_{n+1}(0;\lambda), \cdots, Y_{2n}(0;\lambda)\right)$$

is called Evans' function.

The important aspect of this definition is that $D(\lambda)$ can be constructed analytically.

THEOREM 1 ([**9**] for nerve equations, [**1**] *etc.* for general cases). *$D(\lambda)$ is an analytic function on the connected component of $\mathbb{C}\setminus\sigma_e(L)$ that contains $\{\lambda \mid \operatorname{Re}\lambda > \beta\}$, and its zeros coincide with the eigenvalues of L, and the multiplicity of each zero coincides with the multiplicity of the eigenvalue. Moreover, $D(\lambda)$ can be defined so that it takes real values on the real line. Here, $\sigma_e(L)$ stands for the essential spectrum of L.*

One important implication of this theorem is that for any simple closed curve K in $\mathbb{C}\setminus\sigma_e(L)$, the number of eigenvalues inside K° (counting their multiplicities) is obtained as the winding number of $D(K)$, without finding zeros of $D(\lambda)$, where K° is the domain enclosed by K (Alexander-Gardner-Jones [**1**], Jones [**24**]). The stability index defined in the next subsection is a framework which is defined to extract this property of Evans' function.

2.3. Stability index. The nonautonomous system (7) can be treated as an autonomous system on $\mathbb{C}^{2n}\times[-1,1]$:

$$\left\{\begin{aligned} Y' &= A(\tau,\lambda)Y, \\ \tau' &= \kappa(1-\tau^2), \end{aligned}\right. \tag{9}$$

where $\xi(\tau) = \frac{1}{2\kappa}\log\left(\frac{1+\tau}{1-\tau}\right)$ (*i.e.*, $\tau(\xi) = \frac{e^{2\kappa\xi}-1}{e^{2\kappa\xi}+1}$) and

$$A(\tau,\lambda) = \begin{cases} A(\xi(\tau),\lambda) & (\tau \neq \pm 1), \\ A_\pm(\lambda) & (\tau = \pm 1). \end{cases}$$

Here κ satisfies $0 < \kappa < \frac{a}{2}$ for the a in hypothesis 2. Notice that $(Y_1(\xi;\lambda), \tau(\xi)), \cdots, (Y_{2n}(\xi;\lambda), \tau(\xi))$ are solutions of (9).

Let $Y_1^\pm(\lambda), \cdots, Y_n^\pm(\lambda)$ be the eigenvectors corresponding to the n eigenvalues of $A_\pm(\lambda)$ with positive real part, and let $E_{\lambda,\pm 1} := \operatorname{span}\left\{Y_1^\pm(\lambda), \cdots, Y_n^\pm(\lambda)\right\}$ be the n-dimensional subspace of $\mathbb{C}^{2n}$ spanned by $Y_1^\pm(\lambda), \cdots, Y_n^\pm(\lambda)$. Similarly, let $E_{\lambda,\tau} := \operatorname{span}\{Y_1(\xi(\tau);\lambda), \cdots, Y_n(\xi(\tau);\lambda)\}$; then $E_{\lambda,\tau}$ converges to $E_{\lambda,-1}$ as $\tau \to -1$, *i.e.*, $\xi \to -\infty$. On the other hand, by Lemma 1, $E_{\lambda,\tau}$ converges to $E_{\lambda,+1}$ as $\tau \to +1$ if λ is not an eigenvalue of L. Therefore, $E_\lambda := \bigcup_{\tau\in[-1,+1]} \{\tau\}\times E_{\lambda,\tau}$ forms an n-dimensional vector bundle over $[-1,+1]$. The stability index is obtained from this.

DEFINITION 3. Let $K \subset \{\lambda \mid \operatorname{Re}\lambda > \beta\}\setminus\sigma(L)$ be a simple closed curve. Then $\mathcal{E}(K)$ is an n-dimensional vector bundle over $\mathfrak{K} := (K^\circ \times \{-1\}) \cup (K \times [-1,+1]) \cup$

$(K^\circ \times \{+1\}) \approx \mathbb{S}^2$ defined as follows:

$$\mathcal{E}(K) := \left(\bigcup_{\lambda \in K^\circ} \{(\lambda, -1)\} \times E_{\lambda,-1} \right) \cup \left(\bigcup_{\substack{\lambda \in K \\ \tau \in [-1,+1]}} \{(\lambda, \tau)\} \times E_{\lambda,\tau} \right) \cup \left(\bigcup_{\lambda \in K^\circ} \{(\lambda, +1)\} \times E_{\lambda,+1} \right).$$

THEOREM 2 ([**1**]). *The first Chern number $c_1(\mathcal{E}(K))$ of $\mathcal{E}(K)$ coincides with the number of zeros of $D(\lambda)$ in K° counting their multiplicities. Thus it coincides with the number of eigenvalues of L in K° counting their multiplicities.*

DEFINITION 4. $c_1(\mathcal{E}(K))$ is called the stability index.

The vector bundle $\mathcal{E}(K)$ is the pull-back of the canonical bundle $\Gamma_n(\mathbb{C}^{2n})$ over the complex Graßmannian manifold $G_n(\mathbb{C}^{2n})$ of n-dimensional subspaces in $\mathbb{C}^{2n}$ by the map

$$\mathcal{G}\colon \mathfrak{K} \to G_n(\mathbb{C}^{2n}) : (\lambda, \tau) \mapsto E_{\tau,\lambda}. \tag{10}$$

$c_1(\mathcal{E}(K))$ in Theorem 2 is the image of the generator 1 of $H_2(\mathfrak{K}) \cong \mathbb{Z}$ in $H_2(G_n(\mathbb{C}^{2n})) \cong \mathbb{Z}$ under the homology homomorphism $\mathcal{G}_*\colon H_2(\mathfrak{K}) \to H_2(G_n(\mathbb{C}^{2n}))$ induced by $\mathcal{G}$; in other words, it counts how many times $\mathcal{G}(\mathfrak{K})$ covers $G_n(\mathbb{C}^{2n})$.

The proof depends on the fact that the number of twists E_λ makes as λ rounds on K coincides with the winding number of $D(K)$.

3. Application of the stability index

The subject in this section is applications of the stability index to travelling waves that are constructed by the singular perturbation method or bifurcation.

3.1. Singular perturbation problems. In many cases, the travelling wave under consideration is constructed by the singular perturbation method. For example, the stability analysis of the pulse in a FitzHugh-Nagumo system in Jones [**24**], which introduced the winding number argument that led to the stability index, also used the fact that the pulse is constructed by the singular perturbation method. The basic character of this setting is the splitting of $\mathcal{E}(K)$ into the Whitney sum $\mathcal{E} = \mathcal{E}_f \oplus \mathcal{E}_s$ of the fast bundle $\mathcal{E}_f(K)$ and the slow bundle $\mathcal{E}_s(K)$, which corresponds to the splitting of the system into the fast system and the slow system in the limit of the small parameter tending to 0. In what follows, this is explained with the example of the predator-prey system, which was treated in Gardner-Jones [**17**].

Consider the following system, which is a special case of (1):

$$\begin{cases} u_{1_t} = \varepsilon^2 u_{1xx} + u_1 f_1(u), \\ u_{2_t} = u_{2xx} + u_2 f_2(u), \end{cases} \qquad (t > 0,\ -\infty < x < +\infty), \tag{11}$$

where $u = (u_1, u_2)$ and $0 < \varepsilon \ll 1$ is treated as a small parameter. The nonlinearities $f_1(u)$, $f_2(u)$ are assumed to have steady states $(0,0)$ and $c = (c_1, c_2)$ $(f_1(c) = f_2(c) = 0)$. (Detailed assumptions are not stated here.)

Let $\zeta = x - \varepsilon\theta t$; then the travelling wave equation with velocity $\varepsilon\theta$ becomes

$$\begin{cases} \varepsilon^2 u_{1\zeta\zeta} + \varepsilon\theta u_{1\zeta} + u_1 f_1(u) = 0, \\ \quad u_{2\zeta\zeta} + \varepsilon\theta u_{2\zeta} + u_2 f_2(u) = 0. \end{cases} \tag{12}$$

The following first-order system is called the slow system:

$$(13)\qquad \begin{cases} \varepsilon\dot{U}_1 = V_1, \\ \varepsilon\dot{V}_1 = -\theta V_1 - U_1 f_1(U), \\ \dot{U}_2 = V_2, \\ \dot{V}_2 = -\varepsilon\theta V_2 - U_2 f_2(U), \end{cases} \qquad (\,^{\cdot} = \tfrac{d}{d\zeta}).$$

On the other hand, the system written in $\xi = \zeta/\varepsilon$, $u_i(\xi) = U_i(\varepsilon\xi)$, and $v_i(\xi) = V_i(\varepsilon\xi)$ $(i = 1, 2)$ is called the fast system:

$$(14)\qquad \begin{cases} u_1' = v_1, \\ v_1' = -\theta v_1 - u_1 f_1(u), \\ u_2' = \varepsilon v_2, \\ v_2' = \varepsilon[-\varepsilon\theta v_2 - u_2 f_2(u)], \end{cases} \qquad (\,' = \tfrac{d}{d\xi}).$$

The singular perturbation method deals with the limit as $\varepsilon \downarrow 0$ of these two systems written in different time scales.

Assume that the solution of $U_1 f_1(U) = 0$ is given by $U_1 = 0$ around $U = (U_1, U_2) = (0, 0)$, and is given by $U_1 = p(U_2)$ around $U = (c_1, c_2)$, and let $S_- := \{U_1 = V_1 = 0\}$ and $S_+ := \{U_1 = p(U_2),\ V_1 = 0\} \subset \mathbb{R}^4$. The following system on S_- and S_+ is obtained by letting $\varepsilon = 0$ in (13):

$$(15)\qquad \begin{cases} \dot{U}_2 = V_2, \\ \dot{V}_2 = -g_\pm(U_2), \end{cases}$$

where $g_-(U_2) = U_2 f_2(0, U_2)$ and $g_+(U_2) = U_2 f_2(p(U_2), U_2)$. On the other hand, the following system is obtained by letting $\varepsilon = 0$ in (14), which has sets of equilibria S_- and S_+:

$$(16)\qquad \begin{cases} u_1' = v_1, \\ v_1' = -\theta v_1 - u_1 f_1(u), \\ u_2' = 0, \\ v_2' = 0, \end{cases} \qquad (\,' = \tfrac{d}{d\xi}).$$

Assume that we have a solution X_R of (15) and a solution x_R of (16) with the following properties:

1. $\lim_{\zeta\to-\infty} X_R(\zeta) = 0$ and $\lim_{\zeta\to+\infty} X_R(\zeta) = (c_1, 0, c_2, 0)$.
2. $X_R(\zeta) \in S_-$ for $\zeta < 0$ and $X_R(\zeta) \in S_+$ for $\zeta > 0$. $X_R(\zeta)$ is continuous at $\zeta = 0$ in its U_2 and V_2 components, although it is discontinuous in its U_1 and V_1 components.
3. Let $\lim_{\zeta\to-0} X_R(\zeta) = (0, 0, \alpha, \beta)$ and $\lim_{\zeta\to+0} X_R(\zeta) = (p(\alpha), 0, \alpha, \beta)$. Then

$$\lim_{\xi\to-\infty} x_R(\xi) = (0, 0, \alpha, \beta) \quad \text{and} \quad \lim_{\xi\to+\infty} x_R(\xi) = (p(\alpha), 0, \alpha, \beta).$$

Let $W_- := \{X_R(\zeta) \mid \zeta < 0\}$, $W_F := \{x_R(\xi) \mid -\infty < \xi < +\infty\}$ and $W_+ := \{X_R(\zeta) \mid \zeta > 0\}$; then $\overline{W_- \cup W_F \cup W_+}$ is a curve in $\mathbb{R}^4$ from 0 to $(c_1, 0, c_2, 0)$, and there is a travelling wave from 0 to c with the following property.

THEOREM 3 ([**10**]). (13) *has a solution* $X(\zeta;\varepsilon)$ *for* $0<\varepsilon\ll 1$ *satisfying*

$$\lim_{\zeta\to-\infty} X(\zeta;\varepsilon)=0 \quad \textit{and} \quad \lim_{\zeta\to+\infty} X(\zeta;\varepsilon)=(c_1,0,c_2,0),$$

which stays in some neighborhood of $\overline{W_-\cup W_F\cup W_+}$.

The linearised eigenvalue problem for the travelling wave is also treated in the form of a slow system with the variable ζ and in the form of a fast system with the variable ξ:

$$(17)\quad \begin{cases} \varepsilon\dot{P}_1=Q_1, \\ \varepsilon\dot{Q}_1=-\theta Q_1-[\frac{\partial}{\partial U_1}(U_1f_1(U))-\lambda]P_1-\frac{\partial}{\partial U_2}(U_1f_1(U))\,P_2, \\ \dot{P}_2=Q_2, \\ \dot{Q}_2=-\varepsilon\theta Q_2-\frac{\partial}{\partial U_1}(U_2f_2(U))\,P_1-[\frac{\partial}{\partial U_2}(U_2f_2(U))-\lambda]P_2, \end{cases} \qquad (\,\dot{}=\tfrac{d}{d\zeta}),$$

$$(18)\quad \begin{cases} p_1'=q_1, \\ q_1'=-\theta q_1-[\frac{\partial}{\partial u_1}(u_1f_1(u))-\lambda]p_1-\frac{\partial}{\partial u_2}(u_1f_1(u))\,p_2, \\ p_2'=\varepsilon q_2, \\ q_2'=\varepsilon\left\{-\varepsilon\theta q_2-\frac{\partial}{\partial u_1}(u_2f_2(u))\,p_1-[\frac{\partial}{\partial u_2}(u_2f_2(u))-\lambda]p_2\right\}, \end{cases} \qquad (\,'=\tfrac{d}{d\xi}),$$

where (U_1,U_2) is the U component of $X(\zeta;\varepsilon)$ and (u_1,u_2) is the u component of $x(\xi;\varepsilon):=X(\varepsilon\xi;\varepsilon)$.

Consider the system (17) with $\varepsilon=0$:

$$(19)\quad \begin{cases} [\lambda-\frac{\partial}{\partial U_1}(U_1f_1(U))]P_1=\frac{\partial}{\partial U_2}(U_1f_1(U))\,P_2, \\ Q_1=0, \\ \dot{P}_2=Q_2, \\ \dot{Q}_2=-\frac{\partial}{\partial U_1}(U_2f_2(U))\,P_1-[\frac{\partial}{\partial U_2}(U_2f_2(U))-\lambda]P_2, \end{cases} \qquad (\,\dot{}=\tfrac{d}{d\zeta}),$$

where (U_1,U_2) is the U component of $X_R(\zeta)$. This system has a solution $E_{2R}(\zeta;\lambda)$ with $E_{2R}(\zeta;\lambda)\to 0$ $(\zeta\to-\infty)$, which is continuous at $\zeta=0$. On the other hand, (18) with $\varepsilon=0$, namely

$$(20)\quad \begin{cases} p_1'=q_1, \\ q_1'=-\theta q_1-[\frac{\partial}{\partial u_1}(u_1f_1(u))-\lambda]p_1-\frac{\partial}{\partial u_2}(u_1f_1(u))\,p_2, \quad (\,'=\tfrac{d}{d\xi}) \\ p_2'=0, \\ q_2'=0, \end{cases}$$

has a solution $e_{1R}(\xi;\lambda)$ with $e_{1R}(\xi;\lambda)\to 0$ $(\xi\to-\infty)$, where (u_1,u_2) is the u component of $x_R(\xi)$. Here $E_{2R}(\zeta;\lambda)$ and $e_{1R}(\xi;\lambda)$ are unique up to multiplication by constants. As explained in the previous section, the vector bundle $\mathcal{E}(K)$ is constructed with independent solutions $E_1(\zeta;\lambda;\varepsilon)$ and $E_2(\zeta;\lambda;\varepsilon)$ of (17), which converges to 0 as $\zeta\to-\infty$. The following theorem in Gardner-Jones [**17**] is essential for the choice of $E_1(\zeta;\lambda;\varepsilon)$ and $E_2(\zeta;\lambda;\varepsilon)$.

THEOREM 4. *Let* $e_1(\xi;\lambda;\varepsilon)=E_1(\varepsilon\xi;\lambda;\varepsilon)$. *Then* $E_1(\zeta;\lambda;\varepsilon)$ *and* $E_2(\zeta;\lambda;\varepsilon)$ *can be chosen so that*

$$\lim_{\varepsilon\downarrow 0} e_1(\xi;\lambda;\varepsilon)=e_{1R}(\xi;\lambda) \quad \textit{and} \quad \lim_{\varepsilon\downarrow 0} E_2(\zeta;\lambda;\varepsilon)=E_{2R}(\zeta;\lambda)$$

for $\zeta \neq 0$. *As a result,* $\mathcal{E}(K)$ *can be decomposed into the Whitney sum* $\mathcal{E}(K) = \mathcal{E}_f(K) \oplus \mathcal{E}_s(K)$ *of subbundles* $\mathcal{E}_f(K)$ *and* $\mathcal{E}_s(K)$ *with the fibers* $\mathrm{span}\{E_1(\zeta;\lambda;\varepsilon)\}$ *and* $\mathrm{span}\{E_2(\zeta;\lambda;\varepsilon)\}$.

In [**17**], $c_1(\mathcal{E}_f(K)) = 1$ and $c_1(\mathcal{E}_s(K)) = 0$ are proven from the facts that $\mathcal{E}_f(K)$ is governed by the limiting system (20) and $\mathcal{E}_s(K)$ is governed by (19). Thus the travelling wave $X(\xi;\varepsilon)$ is stable, because the following is obtained by the additivity of the Chern number:

THEOREM 5. $c_1(\mathcal{E}(K)) = c_1(\mathcal{E}_f(K)) + c_1(\mathcal{E}_s(K)) = 1$.

A similar argument is also used in Gardner-Jones [**16**], Miller [**30**], and Kapitula [**25**].

3.2. Bifurcation problems. Another setting in which the stability index is naturally applied is the problem of determining the stability of travelling waves that are generated by bifurcations. In this case, the stability of the original waves is assumed to be known, and the stability of bifurcating waves is considered.

There is a pioneering work by Yanagida-Maginu [**49**] in this context, which treated a double pulse in the nerve equations generated by concatenating two copies of a stable pulse. In [**49**], it is proven that the number of zeros near the origin of the Evans' function for the double pulse is twice that for the original pulse. In what follows, the stability index is applied to prove that this additivity generally holds, even for eigenvalues apart from the origin.

Assume that (1) has three stable steady state solutions $u \equiv u_-, u_0, u_+$ (they do not have to be different), and assume that (4) has two travelling waves $u_1(\xi)$ and $u_2(\xi)$ for $\theta = \Theta$ with the following properties:

1. $\lim_{\xi\to-\infty} u_1(\xi) = u_-$ and $\lim_{\xi\to+\infty} u_1(\xi) = u_0$.
2. $\lim_{\xi\to-\infty} u_2(\xi) = u_0$ and $\lim_{\xi\to+\infty} u_2(\xi) = u_+$.

This is equivalent to (3) having solutions $(u_1(\xi), u_1'(\xi))$ and $(u_2(\xi), u_2'(\xi))$ satisfying the following:

1. $\lim_{\xi\to-\infty} (u_1(\xi), u_1'(\xi)) = (u_-, 0)$ and $\lim_{\xi\to+\infty} (u_1(\xi), u_1'(\xi)) = (u_0, 0)$;
2. $\lim_{\xi\to-\infty} (u_2(\xi), u_2'(\xi)) = (u_0, 0)$ and $\lim_{\xi\to+\infty} (u_2(\xi), u_2'(\xi)) = (u_+, 0)$.

Under this assumption with appropriate conditions, there exists a travelling wave $u(\xi)$, *i.e.*, a solution of (3) $(u(\xi), u'(\xi))$ for $\theta = \theta_0$ near Θ satisfying the following (Figure 2):

$$\lim_{\xi\to-\infty} u(\xi) = u_- \quad \text{and} \quad \lim_{\xi\to+\infty} u(\xi) = u_+,$$

or

$$\lim_{\xi\to-\infty} (u(\xi), u'(\xi)) = (u_-, 0) \quad \text{and} \quad \lim_{\xi\to+\infty} (u(\xi), u'(\xi)) = (u_+, 0).$$

Under this setting, Alexander-Jones [**2**] proved the following additivity.

THEOREM 6. *Let* $\mathcal{E}(K)$ *be the bundle corresponding to* $u(\xi)$, *and let* $\mathcal{E}_1(K)$ *and* $\mathcal{E}_2(K)$ *be the bundles corresponding to* $u_1(\xi)$ *and* $u_2(\xi)$. *Then,*

$$c_1(\mathcal{E}(K)) = c_1(\mathcal{E}_1(K)) + c_1(\mathcal{E}_2(K))$$

if θ_0 *is sufficiently near to* Θ.

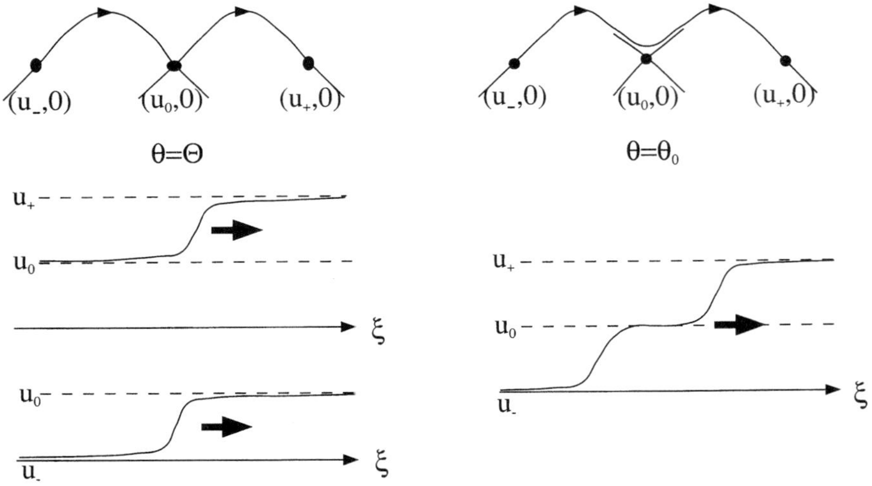

FIGURE 2

The outline of the proof is as follows.

Consider the Bermuda shorts $B := B_0 \cup B_\pm \cup B_\cap$ defined by

$$\begin{aligned} B_0 &= K \times \{(\xi, y) \in (-\infty, \infty) \times [-1, +1] : y > -\xi^2\}, \\ B_\pm &= K^\circ \times \{\pm\infty\} \times [-1, +1], \\ B_\cap &= K^\circ \times \{(\xi, -\xi^2) : -1 \le \xi \le 1\}. \end{aligned}$$

The subset of B for $y = -1$ is homeomorphic to the disjoint union of two copies of $\mathbb{S}^2$, and the subset for $y = +1$ is homeomorphic to $\mathbb{S}^2$. Attach $\mathcal{E}_1(K)$ and $\mathcal{E}_2(K)$ on the two $\mathbb{S}^2$ for $y = -1$, and $\mathcal{E}(K)$ on the $\mathbb{S}^2$ for $y = +1$; then this bundle can be extended to all of B, since $u(\xi)$ is near to the concatenation of $u_1(\xi)$ and $u_2(\xi)$. Therefore the cobordism invariance of the Chern number gives the conclusion. A similar result is proven in Nii [**33**].

It is easy to see from this additivity that $u(\xi)$ is unstable if either of $u_1(\xi)$ or $u_2(\xi)$ is unstable. For example, if $u_1(\xi)$ is unstable, then the linearised operator L_1 along $u_1(\xi)$ has an eigenvalue λ_1 with positive real part. Therefore, for K enclosing λ_1, $\mathcal{E}(K) \ge 1$ follows from $\mathcal{E}_1(K) = 1$ and the additivity. If we take K within the right half-plane, this means that the linearisation L along $u(\xi)$ has an eigenvalue with positive real part, and thus $u(\xi)$ is unstable. This argument is used in the proof of instability in Gardner [**13**].

On the other hand, the additivity does not determine the stability of $u(\xi)$ if $u_1(\xi)$ and $u_2(\xi)$ are both stable. In such a case, the linearised operators L_1 and L_2 have a simple eigenvalue at the origin, and any other eigenvalues are smaller than β (< 0). This means that L has two eigenvalues near the origin, and no other eigenvalue can cause instability of $u(\xi)$. Of course, one of them is at the origin, and thus the sign of the other one determines the stability of $u(\xi)$. (It is easy to see that this eigenvalue is real.)

One way to determine the sign is to determine the sign of the derivative of the Evans' function at the origin: $D'(0)$. If $D(\lambda)$ is normalised so that $D(\lambda) > 0$ when $\lambda \to +\infty$, then $D'(0) > 0$ means that there are an even number of positive eigenvalues. This means that there are no positive eigenvalues, because there can be at most one eigenvalue on the positive part of the real line. On the other hand, $D'(0) < 0$ means that there are an odd number of positive eigenvalues, and thus the eigenvalue under consideration is positive. The following theorem determines the sign of $D'(0)$ from the bifurcation structure:

THEOREM 7. (See [**9**] for the nerve equations, and [**2**], [**29**], etc., for general cases.) *Take the solutions $Y_i(\xi : \lambda)$ of (7) in the definition of $D(\lambda)$ so that*

$$Y_1(\xi : 0) = Y_{n+1}(\xi : 0) = (u'(\xi), u''(\xi)).$$

Let $X_u(\xi;\theta)$ and $X_s(\xi;\theta)$ be one-parameter families of solutions of (3) depending on θ smoothly that satisfy

$$\lim_{\xi \to -\infty} X_u(\xi;\theta) = (u_-, 0) \quad \text{and} \quad \lim_{\xi \to +\infty} X_s(\xi;\theta) = (u_+, 0);$$

moreover, $X_u(\xi;\theta_0) = X_s(\xi;\theta_0) = (u(\xi), u'(\xi))$ for $\theta = \theta_0$. Then $D'(0)$ is expressed as follows:

$$\left.\frac{dD}{d\lambda}\right|_{\lambda=0}$$
$$= \det\left(\left.\frac{\partial X_s}{\partial \theta} - \frac{\partial X_u}{\partial \theta}\right|_{\substack{\theta=\theta_0 \\ \xi=0}}, Y_2(0;\lambda), \cdots, Y_n(0;\lambda), Y_{n+1}(0;\lambda), \cdots, Y_{2n}(0;\lambda)\right).$$

This formula has been applied to concrete examples in Alexander-Jones [**2**], [**3**] and Nii [**33**].

4. Extensions of the stability index

As has been explained, the stability index or, in its original form, the Evans' function, was developed for travelling waves that approach a stable steady state as $\xi \to \pm\infty$. However, it has been extended to differents types of problems. In this section, we discuss applications of the stability index to problems such as boundary value problems in finite intervals and periodic travelling waves.

4.1. Boundary value problems. The first extension was to boundary value problems in finite intervals by Gardner-Jones [**18**].

Consider a parabolic system in a finite interval:

$$\text{(21)} \qquad \begin{cases} u_t = Du_{xx} + f(x, u, u_x), \\ u(x, 0) = u_0(x), \\ B_0 u = 0, \\ B_1 u = 0, \end{cases} \qquad (t > 0,\ 0 < x < 1,\ u(x,t) \in \mathbb{R}^n),$$

where D is a positive diagonal matrix, $f\colon \mathbb{R}^{2n+1} \to \mathbb{R}^n$ is of class C^2, and B_0 and B_1 are the following:

$$\text{(22)} \qquad \begin{aligned} B_0 u &= D^0 u(0,t) + N^0 u_x(0,t), \\ B_1 u &= D^1 u(1,t) + N^1 u_x(1,t). \end{aligned}$$

Here, D^0, D^1, N^0 and N^1 are diagonal matrices with the (i,i) entries α_i^0, α_i^1, β_i^0 and β_i^1 satisfying $(\alpha_i^j)^2 + (\beta_i^j)^2 = 1$ $(1 \le i \le n,\ j = 0, 1)$.

In what follows (21) is assumed to have a stationary solution $\bar{u}(x)$, and the linearisation of (21) at $\bar{u}(x)$ is considered:

$$p_t = Lp := Dp_{xx} + a(x)p + b(x)p_x, \tag{23}$$

where $a(x) = d_u f(x, \bar{u}(x), \bar{u}_x(x))$ and $b(x) = d_{u_x} f(x, \bar{u}(x), \bar{u}_x(x))$. Letting

$$Y = \begin{pmatrix} p \\ q \end{pmatrix} \in \mathbb{C}^{2n} \quad \text{and} \quad A(x,\lambda) = \begin{pmatrix} 0 & I \\ D^{-1}(\lambda I - a(x)) & -D^{-1}b(x) \end{pmatrix},$$

the eigenvalue problem $Lp = \lambda p$ of L is treated in the following form:

$$Y' = A(x,\lambda)Y, \qquad Y(0) \in U_0,\ Y(1) \in U_1, \tag{24}$$

where $U_j := \{(p,q) \in \mathbb{C}^{2n} \mid \alpha_i^j p_i + \beta_i^j q_i = 0\ (1 \le i \le n)\}$ $(j = 0,1)$ is an n-dimensional subspace in $\mathbb{C}^{2n}$.

Let $u_i^j = (0,\cdots,0,\beta_i^j,0,\cdots,0,-\alpha_i^j,0,\cdots,0)$ $(1 \le i \le n)$ be the basis of U_j $(j = 0,1)$, and let $v_i^j = (0,\cdots,0,\alpha_i^j,0,\cdots,0,\beta_i^j,0,\cdots,0)$ $(1 \le i \le n)$ be the basis of $V_j := U_j^\perp$. Let θ_i^j be the angle between (α_i^j, β_i^j) and $(1,0)$ on the plane, and let $R(\theta)$ be the rotation of the angle θ on the plane. If

$$(\alpha_i^x, \beta_i^x) = R\left(x(\theta_i^1 - \theta_i^0)\right)(\alpha_i^0, \beta_i^0),$$

then the subspace U_x spanned by $u_i^x := (0,\cdots,0,\beta_i^x,0,\cdots,0,-\alpha_i^x,0,\cdots,0)$ and the subspace V_x spanned by $v_i^x := (0,\cdots,0,\alpha_i^x,0,\cdots,0,\beta_i^x,0,\cdots,0)$ are both n-dimensional, and $V_x := U_x^\perp$. Moreover, let $\varphi_i(x;\lambda)$ $(1 \le i \le n)$ be a solution of (24) satisfying $\varphi_i(0;\lambda) = u_i^0$, and let $\Phi(x;\lambda)$ be the subspace of $\mathbb{C}^{2n}$ spanned by $\{\varphi_i(x;\lambda)\}$.

In this notation, the vector bundle $\mathcal{E}(K)$ over $\mathfrak{K}' := (K^\circ \times \{0\}) \cup (K \times [0,1]) \cup (K^\circ \times \{1\}) \approx \mathbb{S}^2$ for a simple closed curve K with $K \cap \sigma(L) = \emptyset$ is constructed as follows.

First, let $A_x := \mathbb{C}^{2n}/U_x \times \mathbb{C}^{2n}/V_x$ $(0 \le x \le 1)$, and consider the following $2n$-dimensional vector bundle:

$$\mathcal{A}(K) := \left(\bigcup_{\lambda \in K^\circ} \{(\lambda,0)\} \times A_0\right) \cup \left(\bigcup_{\substack{\lambda \in K \\ x \in [0,1]}} \{(\lambda,x)\} \times A_x\right) \cup \left(\bigcup_{\lambda \in K^\circ} \{(\lambda,1)\} \times A_1\right).$$

Let $E_{\lambda,0} := \{\bar{0}\} \times \mathbb{C}^{2n}/V_0$ and $E_{\lambda,1} := \mathbb{C}^{2n}/U_1 \times \{\bar{0}\}$, with $\{\bar{0}\}$ standing for the equivalence class of $0 \in \mathbb{C}^{2n}$ in $\mathbb{C}^{2n}/U_0$ or in $\mathbb{C}^{2n}/V_1$, and let $E_{\lambda,x}$ be the subspace of A_x spanned by $\chi_i(x;\lambda) := (\varphi_i(x;\lambda) + U_x, (1-x)\varphi_i(x;\lambda) + V_x)$. In this setting, $\mathcal{E}(K)$, defined below, becomes an n-dimensional subbundle of $\mathcal{A}(K)$, because $\Phi(1;\lambda) \cap U_1 = \emptyset$ if $\lambda \notin \sigma(L)$:

$$\mathcal{E}(K) := \left(\bigcup_{\lambda \in K^\circ} \{(\lambda,0)\} \times E_{\lambda,0}\right) \cup \left(\bigcup_{\substack{\lambda \in K \\ x \in [0,1]}} \{(\lambda,x)\} \times E_{\lambda,x}\right) \cup \left(\bigcup_{\lambda \in K^\circ} \{(\lambda,1)\} \times E_{\lambda,1}\right).$$

The following holds for $\mathcal{E}(K)$.

THEOREM 8 ([**18**]). *Let K be a simple closed curve with $K \cap \sigma(L) = \emptyset$. Then the number of eigenvalues of L in the domain K° enclosed by K is equal to the first Chern number $c_1(\mathcal{E}(K))$ of $\mathcal{E}(K)$.*

Gardner [**11**] applied this framework to a concrete example, where a singular perturbation problem is treated, and the bundle $\mathcal{E}(K)$ splits into subbundles $\mathcal{E}(K) = \mathcal{E}_f(K) \oplus \mathcal{E}_s(K)$ as before.

4.2. Periodic travelling waves. In this section, the linearised eigenvalue problem along periodic travelling waves—*i.e.*, a spatially periodic stationary solution of (2)—is treated. This problem is different from the previous problems in the following sense. The spectra appearing in the stability problem of periodic travelling waves are essential spectra, and characteristics of the spectra themselves have to be investigated before the stability index is defined. Gardner [**12**] defined the notion of γ-eigenvalue and characterized the essential spectra, and defined the stability index for them.

Assume that (2) has a spatially periodic solution $u(\xi)$ with period T, and consider a linearised eigenvalue problem along $u(\xi)$:

$$Lp = Bp_{\xi\xi} + \theta p_\xi + Df\,(u(\xi))\,p = \lambda p. \tag{25}$$

By letting

$$Y = \begin{pmatrix} p \\ q \end{pmatrix} \in \mathbb{C}^{2n} \quad \text{and} \quad A(\xi;\lambda) := \begin{pmatrix} 0 & I \\ B^{-1}\left(\lambda I - Df\,(u(\xi))\right) & -\theta B^{-1} \end{pmatrix},$$

this system is also treated as the following first-order system:

$$Y' = A(\xi;\lambda)Y. \tag{26}$$

Then the notion of a γ-eigenvalue is defined as follows.

DEFINITION 5 ([**12**]). Let γ be a complex number with $|\gamma| = 1$. Then λ is a γ-eigenvalue of L if and only if (26) has a solution satisfying the boundary condition $Y(T;\lambda) = \gamma Y(0;\lambda)$; that is, for the fundamental matrix solution $S(\xi;\lambda)$ of (26) with $S(0;\lambda) = I$, $S(T;\lambda)$ has the eigenvalue γ.

λ's being a γ-eigenvalue for some γ is equivalent, of course, to (25)'s having a nontrivial bounded solution. Under this definition, γ-eigenvalues become discrete for each γ. Thus the stability index, which counts the number of γ-eigenvalues, is defined.

Consider a coupled system of (26) and a trivial equation $W' = 0$:

$$\begin{cases} Y' = A(\xi;\lambda)Y, \\ W' = 0. \end{cases} \tag{27}$$

Let $X_\gamma = \{(\gamma Y, Y) \in \mathbb{C}^{4n} \mid Y \in \mathbb{C}^{2n}\}$; then λ is a γ-eigenvalue if and only if (27) has a solution $Z(\xi) = (Y(\xi), W(\xi))$ with $Z(0) \in X_1$ and $Z(T) \in X_\gamma$. Let $Z_i(\xi;\lambda)$ be the solution of (27) satisfying $Z_i(0;\lambda) = (e_i, e_i) \in X_1$ for the basis $\{e_i\}$ of $\mathbb{C}^{2n}$, and let $\Psi(\xi;\lambda)$ be the subspace of $\mathbb{C}^{4n}$ spanned by $\{Z_i(\xi;\lambda)\}$. Fix a curve $\gamma(\xi)$ $(0 \le \xi \le T)$ on $\mathbb{C}$ satisfying $|\gamma(\xi)| = 1$, $\gamma(0) = 1$ and $\gamma(T) = \gamma$.

In this notation, for a simple closed curve K on the complex plane that does not intersect with the set of γ-eigenvalues of L, a vector bundle $\mathcal{E}(K;\gamma)$ over $\mathfrak{K}'' := (K^\circ \times \{0\}) \cup (K \times [0,T]) \cup (K^\circ \times \{T\}) \approx \mathbb{S}^2$ is constructed. First, let

$$A_\xi := \mathbb{C}^{4n}/X_{\gamma(\xi)} \times \mathbb{C}^{4n}/X^\perp_{\gamma(\xi)} \qquad (0 \le \xi \le T)$$

and consider the following $4n$-dimensional vector bundle:

$$\mathcal{A}(K) := \left(\bigcup_{\lambda \in K^\circ} \{(\lambda, 0)\} \times A_0 \right) \cup \left(\bigcup_{\substack{\lambda \in K \\ \xi \in [0,T]}} \{(\lambda, \xi)\} \times A_\xi \right) \cup \left(\bigcup_{\lambda \in K^\circ} \{(\lambda, T)\} \times A_T \right).$$

Let $E_{\lambda,0} := \{\bar{0}\} \times \mathbb{C}^{4n} / X_1^\perp$ and $E_{\lambda,T} := \mathbb{C}^{4n} / X_\gamma \times \{\bar{0}\}$, and let $E_{\lambda,\xi}$ be the subspace of A_ξ spanned by $\zeta_i(\xi;\lambda) := \left(Z_i(\xi;\lambda) + X_{\gamma(\xi)}, \frac{T-\xi}{T} Z_i(\xi;\lambda) + X_{\gamma(\xi)}^\perp \right)$. Then $\mathcal{E}(K;\gamma)$ below is a $2n$-dimensional subbundle of $\mathcal{A}(K)$, since $\Psi(T;\lambda) \cap X_\gamma = \emptyset$ if λ is not a γ-eigenvalue:

$$\mathcal{E}(K;\gamma) := \left(\bigcup_{\lambda \in K^\circ} \{(\lambda, 0)\} \times E_{\lambda,0} \right) \cup \left(\bigcup_{\substack{\lambda \in K \\ \xi \in [0,T]}} \{(\lambda, \xi)\} \times E_{\lambda,\xi} \right) \cup \left(\bigcup_{\lambda \in K^\circ} \{(\lambda, T)\} \times E_{\lambda,T} \right).$$

THEOREM 9 ([**12**]). *For a simple closed curve K on the complex plane that does not intersect the set of γ-eigenvalues of L, the number of γ-eigenvalues of L in the domain K° enclosed by K is equal to the first Chern number $c_1(\mathcal{E}(K;\gamma))$ of $\mathcal{E}(K;\gamma)$.*

As for the spectrum $\sigma(L)$ itself, this theorem, for example, has the following implication.

COROLLARY 1. *If $c_1(\mathcal{E}(K;\gamma)) = 1$ for a simple closed curve K in $\mathbb{C} \setminus \sigma(L)$, then $\sigma(L) \cap K^\circ$ is an image of $\mathbb{S}^1$ under a continuous map.*

This stability index is used for the proof of the instability of Ginzburg-Landau almost periodic travelling waves in Doelman-Gardner-Jones [**5**].

REMARK 4. There is necessarily a 1-eigenvalue at the origin for a periodic travelling wave, and thus the essential spectrum includes the origin. Therefore, the general theory of infinite-dimensional dynamical systems is not sufficient for stability.

4.3. The index restricted to real eigenvalues. The previous section dealt with the extension of the stability index to problems that were not in the original framework. This section deals with the extension to problems for which the original form of the index is applicable but not sufficient.

Theorem 6 in section 3.2 holds for general cases where $u(\xi)$ is the concatenations of k ($k \geq 2$) waves $u_1(\xi), \cdots, u_k(\xi)$, that is,

$$c_1(\mathcal{E}(K)) = \sum_{i=1}^{k} c_1(\mathcal{E}_i(K)).$$

This implies that if $u_1(\xi), \cdots, u_k(\xi)$ are all stable, then for the linearisation L along $u(\xi)$, there are $k-1$ eigenvalues near the origin that can destabilise $u(\xi)$.

$D'(0) < 0$ again means the existence of an odd number of positive eigenvalues as in section 3.2, and instability follows. On the other hand, $D'(0) > 0$ does not mean existence nor nonexistence of eigenvalues. This is the limitation of the stability index which comes from the fact that $\mathcal{E}(K)$ is constructed for the simple closed curve K on the complex plane and only detects the number of eigenvalues in K°, not their ordering. This problem was solved for a concrete example in Nii [**34**] and [**35**]. In particular, in [**35**] the ordering of eigenvalues is determined by restricting the index to real eigenvalues.

Let $\beta < \lambda_1 < \lambda_2$ be real numbers, and restrict the domain of definition of $\mathcal{G}$ in (10) to $\mathfrak{k} := ([\lambda_1, \lambda_2] \times \{-1\}) \cup (\{\lambda_1, \lambda_2\} \times [-1, +1]) \cup ([\lambda_1, \lambda_2] \times \{+1\}) \approx \mathbb{S}^1$. Regard $E_{\lambda,\tau}$ as an n-dimensional subspace of $\mathbb{R}^{2n}$ for $\lambda \in [\lambda_1, \lambda_2]$; then $\mathcal{G}|_{\mathfrak{k}}$ can be regarded as a map from $\mathfrak{k}$ to the real Graßmannian manifold $G_n(\mathbb{R}^{2n})$. Moreover, if $\mathcal{G}|_{\mathfrak{k}}(\mathfrak{k})$ is included in a subset Ω of $G_n(\mathbb{R}^{2n})$ that is homotopic to $\mathbb{S}^1$, then the number of eigenvalues in $[\lambda_1, \lambda_2]$ can be detected by looking at the image of the generator 1 of $H_1(\mathfrak{k}) \cong \mathbb{Z}$ in $H_1(\Omega) \cong \mathbb{Z}$ under the homomorphism $\mathcal{G}|_{\mathfrak{k}*}\colon H_1(\mathfrak{k}) \to H_1(\Omega)$ induced by $\mathcal{G}|_{\mathfrak{k}}$. In [**35**], the stability of $u(\xi)$ is proven by constructing Ω above, exploiting two-dimensionality of the system, and finding $k-1$ eigenvalues in $[\lambda_1, \lambda_2]$ for some $\lambda_1 < \lambda_2 < 0$.

5. Advantages and limitations of the stability index

In this section, advantages and limitations of the stability index are discussed.

First of all, one advantage of the stability index is that it depends on geometric qualitative information about the system, and does not require detailed analytical information.

The treatment of heteroclinic bifurcations of travelling waves in Nii [**33**] provides a good example. That is, when determining the stability of a travelling wave that is generated by what is called a gluing bifurcation from two stable travelling waves, information to determine the bifurcation is also sufficient to determine the stability of the wave, and thus merely the bifurcation diagram is enough for some cases. This is quite different from the argument in Ikeda [**21**], [**22**], in which the pulses are reconstructed by the singular perturbation method, although their existence has been proven in Kokubu-Nishiura-Oka [**27**] by bifurcation, or what is called Lin's method introduced in Sandstede [**42**], in which the bifurcation itself has to be treated by Lin's method in order to obtain sufficient analytical information.

Moreover, in Nii [**33**], its qualitativity allows us to define the stability index for orbits near to but not on a heteroclinic orbit.

On the other hand, as a limitation of the dependence on qualitative information, it is almost impossible to obtain detailed information on dependence of the eigenvalues on the small parameters or bifurcation parameters. In this aspect, the SLEP method for singular perturbation problems or Lin's method for bifurcation problems has an advantage.

Another disadvantage of the stability index is that it is a framework to deal with only eigenvalues and does not give any information such as the profile of eigenfunctions. Therefore, the instability mechanism for unstable travelling waves cannot be obtained.

Lin's method, mentioned above for the sake of comparison with the stability index, is the method originally developed to treat bifurcation of vector fields, and was extended by Sandstede [**42**] to treat linearised eigenvalue problems for bifurcating travelling waves. The basic framework is to reduce the eigenvalue problem around 0 to the eigenvalue problem of a matrix that represents the dominant term of the perturbation from $\lambda = 0$. This approach has also been successfully applied to many problems (Sandstede [**42**], [**43**], Sandstede-Jones-Alexander [**44**]), and the relation between Lin's method and Evans' function is mentioned in Sandstede [**42**].

On the other hand, the SLEP (= Singular Limit Eigenvalue Problem) method is the method developed to treat singular perturbation problems, and it is also successful (Nishiura [**36**], [**37**], Nishiura-Fujii [**38**], Nishiura-Mimura-Ikeda-Fujii [**39**], and so forth). The basic idea is to obtain information as the limit small parameter $\varepsilon \downarrow 0$ by properly scaling the eigenvalue problem. Ikeda-Nishiura-Suzuki [**23**] proved that the SLEP function coincides with the dominant term of Evans' function in the limit as $\varepsilon \downarrow 0$.

6. Conclusion

This article has explained briefly a topological method for linearised eigenvalue problems along travelling waves introduced by Alexander-Gardner-Jones [**1**]. One recent development not mentioned is that Doelman-Gardner-Kaper [**14**], Gardner-Zumbrum [**15**], and Kapitula-Sandstede [**26**] extended the domain of definition of Evans' function into the essential spectrum.

In section 2.2 and section 2.3, Evans' function $D(\lambda)$ or the vector bundle $\mathcal{E}(K)$ are constructed with the solutions $Y_1(\xi;\lambda), \cdots, Y_n(\xi;\lambda)$ of (7) that decay to 0 as $\xi \to -\infty$. If λ is in the essential spectrum of L, however, some of $Y_1(\xi;\lambda), \cdots, Y_n(\xi;\lambda)$ do not converge to 0 when $\xi \to -\infty$, and some of $Y_{n+1}(\xi;\lambda), \cdots, Y_{2n}(\xi;\lambda)$ do not converge to 0 when $\xi \to +\infty$. When Evans' function $D(\lambda)$ is constructed regardless of this fact, $D(\lambda)$ becomes multivalued, and has to be considered as a function on a proper Riemannian surface. Using this extension, Doelman-Gardner-Kapper [**14**] treated the case where an unstable travelling wave of Fisher type appears in the construction by singular perturbation, and constructed the bundle $\mathcal{E}(K)$, and calculated $c_1(\mathcal{E}(K))$. Kapitula-Sandstede [**26**] also treated similar extensions of Evans' function and solitary waves in a nonlinear Schrödinger equation.

Rubin [**41**] also extended the stability index to a stationary solution of spatially inhomogeneous systems in an infinite interval, by regarding the system as a perturbation from a spatially homogeneous system.

The following are some of the remaining problems.

First, the geometric characterisation of the case where all the eigenvalues are real is not known. Nii [**35**] proved the nonexistence of eigenvalues with positive real part by finding all $k-1$ eigenvalues on the negative real line. However, the reason why all the eigenvalues are real is not clear. On the other hand, the stability can be shown also by Lin's method (Sandstede [**43**]), but realness of the eigenvalues cannot be shown in that case. Sandstede-Jones-Alexander [**44**] proved that all eigenvalues are real in a reversible system by Lin's method, but the geometric reason for this case also is not known.

In some examples, relations between the distribution of the eigenvalues and the twisting of an unstable manifold of equilibrium $(u_-, 0)$ or a stable manifold of $(u_+, 0)$ of the system (3) appear, and some geometric explanation is hoped for (Sandstede-Nii [**45**]), but is not yet known.

7. A remark on the English translation

The original Japanese version of this article was written based on the information available to the author in 1997, and much progress has been made since then. Readers interested in this topic are recommended to see a review article on Evans' function by R. Gardner, T. Kapitula and C. Jones, which is now in preparation.

References

[1] J. Alexander, R. Gardner, and C. Jones, *A topological invariant arising in the stability analysis of travelling waves.* J. Reine Angew. Math. **410** (1990), pp. 167–212.

[2] J. Alexander and C. Jones, *Existence and stability of asymptotically oscillatory double pulses.* J. Reine Angew. Math. **446** (1994), pp. 49–79.

[3] ———, *Existence and stability of asymptotically oscillatory triple pulses.* Z. Angew. Math. Phys. **44** (1993), pp. 189–200.

[4] P. Bates and C. Jones, *Invariant manifolds for semilinear partial differential equations.* Dynamics Reported **2** (1988), pp. 1–38.

[5] A. Doelman, R. Gardner, and C. Jones, *Instability of quasiperiodic solutions of the Ginzburg-Landau equation.* Proc. Roy. Soc. Edinburgh Sect. A **125** (1995), pp. 501–517.

[6] J. Evans, *Nerve axon equations, I: Linear approximations.* Indiana Univ. Math. J. **21** (1972), pp. 877–955.

[7] ———, *Nerve axon equations, II: Stability at rest.* Indiana Univ. Math. J. **22** (1972), pp. 75–90.

[8] ———, *Nerve axon equations, III: Stability of the nerve impulse.* Indiana Univ. Math. J. **22** (1972), pp. 577–594.

[9] ———, *Nerve axon equations, IV: The stable and unstable impulses.* Indiana Univ. Math. J. **24** (1975), pp. 1169–1190.

[10] R. Gardner, *Topological methods arising in the study of travelling waves.* Reaction-Diffusion Equations (K. J. Brown and A. A. Lacey, eds.), Clarendon Press, Oxford, 1990, pp. 173–198.

[11] ———, *Stability and Hopf bifurcation of steady state solutions of a singularly perturbed reaction-diffusion system.* SIAM J. Math. Anal. **23** (1992), pp. 99–149.

[12] ———, *On the structure of the spectra of periodic travelling waves.* J. Math. Pures Appl. **72** (1993), pp. 415–439.

[13] ———, *Instability of oscillatory shock profile solutions of the generalized Burgers-KdV equation.* Physica D **90** (1996), pp. 366–386.

[14] A. Doelman, R. Gardner, and T. Kaper, *A stability index analysis of 1-D patterns of the Gray-Scott model*, preprint.

[15] R. Gardner and K. Zumbrum, *The gap lemma and geometric criteria for instability of viscous shock profiles.*, Comm. Pure Appl. Math. **51** (1998), pp. 789-847.

[16] R. Gardner and C. Jones, *Travelling waves of a perturbed diffusion equation arising a phase field model.* Indiana Univ. Math. J. **38** (1989), pp. 1197–1222.

[17] ———, *Stability of travelling wave solutions of diffusive predator-prey systems.* Trans. Amer. Math. Soc. **327** (1991), pp. 465–524.

[18] ———, *A stability index for steady state solutions of boundary value problems for parabolic systems.* J. Differential Equations **91** (1991), pp. 181–203.

[19] D. Henry, *The geometric theory of semilinear parabolic equations.* Lecture Notes in Math. **840**, Springer, 1981.

[20] H. Ikeda, *Singular perturbation approach to stability properties of travelling wave solutions of reaction-diffusion systems.* Hiroshima Math. J. **19** (1989), pp. 587–630.

[21] ———, *Singular pulse wave bifurcations from front and back waves in bistable reaction-diffusion-systems.* Methods Appl. Anal. **3** (1996), pp. 285–317.

[22] ———, *Existence and stability of pulse waves bifurcated from front and back waves in reaction-diffusion systems.* Japan. J. Indust. Appl. Math. **15** (1998), pp. 163–231.

[23] H. Ikeda, Y. Nishiura, and H. Suzuki, *Stability of travelling waves and a relation between the Evans function and SLEP equation.* J. Reine Angew. Math. **475** (1996), pp. 1–37.

[24] C. Jones, *Stability of the travelling wave solution of the FitzHugh-Nagumo system.* Trans. Amer. Math. Soc. **286** (1984), pp. 431–469.

[25] T. Kapitula, *Existence and stability of singular heteroclinic orbits for the Ginzburg-Landau equation.* Nonlinearity **9** (1996), pp. 669–685.

[26] T. Kapitula and B. Sandstede, *Stability of bright solitary-wave solutions to perturbed non-linear Schrödinger equations.* Physica D **124** (1998), pp. 58–103.

[27] H. Kokubu, Y. Nishiura, and H. Oka, *Heteroclinic and Homoclinic bifurcations in bistable reaction diffusion systems.* J. Differential Equations **86** (1990), pp. 260–341.

[28] X. Lin, *Using Melnikov's method to solve Silnikov's problems.* Proc. Roy. Soc. Edinburgh Sect. A **116** (1990), pp. 295–325.

[29] K. Maginu, *A geometric condition for the instability of solitary travelling wave solutions in reaction-diffusion equation.* preprint

[30] P. Miller, *Stability of travelling waves for three-componemt reaction-diffusion system.* Ph.D. thesis, University of Massachusetts, Amherst, 1994.

[31] J. Milnor and J. Stasheff, *Characteristic Classes.* Princeton Univ. Press, Princeton, NJ, 1974.

[32] J. Murray, *Mathematical Biology.* Second, corrected edition, Springer-Verlag, Berlin, 1993.

[33] S. Nii, *An extension of the stability index for traveling-wave solutions and its application to bifurcations.* SIAM J. Math. Anal. **28** (1997), pp. 402–433.

[34] ———, *Stability of travelling multiple-front (multiple-back) wave solutions of the FitzHugh-Nagumo equations.* SIAM J. Math. Anal. **28** (1997), pp. 1094–1112.

[35] ———, *A topological proof of stability of N-front solutions of the FitzHugh-Nagumo equations.* J. Dynam. Differential Equations **11** (1999), pp. 515–555.

[36] Y. Nishiura, *Singular limit approach to stability and bifurcation for bistable reaction diffusion systems.* Rocky Mountain J. Math. **21** (1991), pp. 727-767.

[37] ———, *Coexistence of infinitely many stable solutions to reaction diffusion systems in the singular limit.* Dynamics Reported (New Series) **3** (1994), Springer.

[38] Y. Nishiura and Fujii, *Stability of singular perturbed solutions to systems of reaction diffusion equations.* SIAM J. Math. Anal. **18** (1987), pp. 1726–1770.

[39] Y. Nishiura, M. Mimura, H. Ikeda, and H. Fujii, *Singular limit analysis of stability of traveling wave solutions in bistable reaction-diffusion systems.* SIAM. J. Math. Anal. **21** (1990), pp. 85–122.

[40] R. Pego and M. Weinstein, *Eigenvalues, and instabilities of solitary waves.* Philosophical Transactions of Royal Society of London **340** (1992), pp. 47–94.

[41] J. Rubin, *The Generation of Edge Oscillations in an Inhomogeneous Reaction-Diffusion System.* Ph.D. thesis, Brown University, 1996.

[42] B. Sandstede, *Stability of multiple-pulse solutions.* Trans. Amer. Math. Soc. **350** (1998), pp. 429–472.

[43] ———, *Stability of N-fronts bifurcating from a twisted heteroclinic loop and an application to the FitzHugh-Nagumo equation.* SIAM J. Math. Anal. **29** (1998), pp. 183–207.

[44] B. Sandstede, C. Jones, and J. Alexander, *Existence and stability of N-pulses on optical fibers with phase-sensitive amplifiers.* Physica D **106** (1997), pp. 167–206.

[45] B. Sandstede and S. Nii, in preparation.

[46] N. Steenrod, *The Topology of Fibre Bundles.* Princeton Univ. Press, Princeton, NJ, 1951.

[47] I. Tamura, *Differential Topology.* I–III, 2nd ed., Iwanami Shoten, Tokyo, 1983 (in Japanese)

[48] E. Yanagida, *Stability of fast travelling pulse solutions of the FitzHugh-Nagumo equations.* J. Math. Biology **22** (1985), pp. 81–104.

[49] E. Yanagida and K. Maginu, *Stability of double-pulse solutions in nerve axon equations.* SIAM J. Appl. Math. **49** (1989), pp. 1158–1173.

Department of Mathematics, Faculty of Science, Saitama University, Urawa City, 338-8570, Japan

Translated by SHUNSAKU NII

Amer. Math. Soc. Transl.
(2) Vol. **204**, 2001

Extension of Almost Periodic Functions and Analyticity on Flows

Jun-ichi Tanaka

1. Introduction

Let D be the unit disc $\{|z| < 1|\}$ in the complex plane $\mathbf{C}$, and let z_0 be a point on the unit circle $\mathbf{T}$. Suppose that $w = f(z)$ is an analytic function on D. It is then important to investigate the distribution of values as well as the behavior of $f(z)$ around the boundary point z_0. Awareness of such subjects is old in function theory, and it seems to be an origin of value distribution theory and the theory of cluster sets.

In the 1940s, I. M. Gelfand and his coworkers built a theory of commutative Banach algebras. As one branch, theories of function algebras began to develop in the early 1950s, which formed a new link between classical function theory and functional analysis. Thus, for example, the theory of Hardy spaces has been extended to an abstract setting.

In this note, by considering the uniform algebras induced by flows, the boundary behavior of $f(z)$ is investigated in the region between the two circles tangent to $\mathbf{T}$ at z_0, although a certain growth restriction of $|f(z)|$ is required by the Phragmén-Lidelöf principle.

The outline of our method runs as follows. Suppose, for simplicity, that $f(z)$ is bounded in D and $z_0 = 1$. Let $z(w)$ be the conformal map of the upper half-plane $\mathbf{R}^2_+$ onto D by

$$z(w) = \frac{w-i}{w+i}.$$

We then put $g(w) = f(z(w))$. Since $z(w)$ maps ∞ to 1, we want to look into the behavior of $g(w)$ around infinity. Let $\beta\mathbf{Z}$ be the Stone-Čech compactification of the integer group $\mathbf{Z}$. Divide $\mathbf{R}^2_+$ into the countably many strips $[n, n+1) \times (0, \infty)$ for n in $\mathbf{Z}$. Since $\{g(n+w); n \in \mathbf{Z}\}$ is a bounded sequence for each ω in $[0,1) \times (0, \infty)$,

This article originally appeared in Japanese in Sūgaku **51** (2) (1999), 113–128.
2000 *Mathematics Subject Classification.* Primary 43A17, 46J10, 46J15, 46J20; Secondary 11K70, 11M06, 28D15, 30H05.
Key words and phrases. Uniform algebras, Hardy spaces, outer functions, minimal flows, Dirichlet series.
This research was partially supported by Grant 10640180 from the Japanese Ministry of Education.

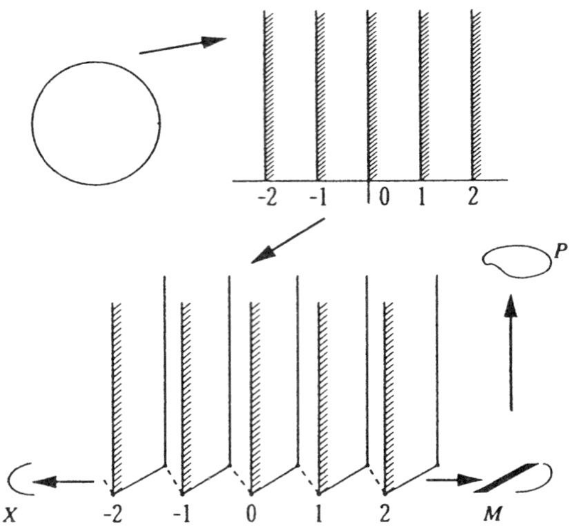

FIGURE 1

$g(w)$ extends to a continuous function on $\beta\mathbf{Z} \times [0,1) \times (0,\infty)$, which represents concretely a portion of the cluster set of $g(w)$. The shift operator $S_0 n = n+1$ on $\mathbf{Z}$ extends to a homeomorphism S on $\beta\mathbf{Z}$. We denote by $\mathbf{X}$ the quotient space obtained from $\beta\mathbf{Z} \times [0,1]$ by identifying $(y,1)$ with $(Sy,0)$. By regarding the real line $\mathbf{R}$ as $\mathbf{Z} \times [0,1]$, the translation on $\mathbf{R}$ induces a continuous flow on $\mathbf{X}$. Thus, using ergodic theory, we try to study the behavior of $g(w)$ on the growth $\mathbf{X} \setminus \mathbf{R}$ (see Figure 1).

In the next section we establish the notation and discuss briefly the principal techniques of Hardy spaces in certain settings. Section 3 treats the uniform algebras induced by flows and, as an application, we deal with a negative answer to a question posed by Forelli [**10**]:

Is the uniform algebra induced by a minimal flow a Dirichlet algebra?

Restricting our attention to the case of almost periodic flows, we discuss in Section 4 analytic functions on a quotient of the Bohr group. After preparing some lemmas, we study Dirichlet series obtained as the limits of translations of the Riemann zeta-function in Section 5, which contains some results published here for the first time.

2. Rudiments of Hardy spaces

We deal with certain subspaces of analytic functions with growth conditions, together with some basic results on boundary behavior. Let $1 \le p < \infty$. A function $f(z)$ analytic in D is said to belong to the *Hardy space* $H^p(D)$ if

$$\|f\|_p^p \;=\; \sup_{0<r<1} \frac{1}{2\pi} \int_0^{2\pi} |f(re^{i\theta})|^p \, d\theta \;<\; \infty,$$

while $H^\infty(D)$ is defined to be the space of all bounded analytic functions $f(z)$ on D, normed by

$$\|f\|_\infty \;=\; \sup\{|f(z)|\,;\, z \in D\}.$$

Let $H^p(d\theta/2\pi)$, $1 \le p \le \infty$, be the space of all functions $f(\theta)$ in $L^p(d\theta/2\pi)$ that have Fourier series of the form

$$f(\theta) \sim \sum_{n=0}^{\infty} a_n e^{in\theta} .$$

Fatou's theorem then enables us to identify $H^p(D)$ and $H^p(d\theta/2\pi)$. The *disc algebra* $A(\mathbf{T})$ is the algebra of all continuous functions in $H^\infty(d\theta/2\pi)$, which is a commutative Banach algebra with the uniform norm. Then $H^p(d\theta/2\pi)$, $1 \le p < \infty$, is also obtained as the closure of $A(\mathbf{T})$ in $L^p(d\theta/2\pi)$, while $H^\infty(d\theta/2\pi)$ is the weak-* closure of $A(\mathbf{T})$ in $L^\infty(d\theta/2\pi)$.

The *Hardy space* $H^p(\mathbf{R}^2_+)$, $1 \le p < \infty$, on $\mathbf{R}^2_+$ is defined similarly to be the space of all analytic functions on $\mathbf{R}^2_+$ such that

$$\|f\|_p^p \;=\; \sup_{0<y<\infty} \int_{-\infty}^{\infty} |f(x+iy)|^p \, dy \;<\; \infty,$$

while $H^\infty(\mathbf{R}^2_+)$ is the space of all bounded analytic functions on $\mathbf{R}^2_+$. We denote by $H^p(dt)$ the subspace of $L^p(dt)$ of all the boundary-value functions of $H^p(\mathbf{R}^2_+)$, and identify $H^p(\mathbf{R}^2_+)$ with $H^p(dt)$ as usual. A function f in $L^p(dt)$ lies in $H^p(dt)$ if and only if the Fourier transform of f vanishes on $(-\infty, 0)$, in a suitable sense. There is a simple relation between $H^p(D)$ and $H^p(\mathbf{R}^2_+)$. Each function g in $H^p(D)$ is transformed by the linear fractional map $z = (w-i)/(w+i)$ into an analytic function f on $\mathbf{R}^2_+$. Since $t = (e^{i\theta} - i)/(e^{i\theta} + i)$ implies

$$\frac{1}{2\pi}\, d\theta \;=\; \frac{1}{\pi}\, \frac{dt}{1+t^2},$$

we denote by $H^p(dt/\pi(1+t^2))$ the space of all the boundary-value functions of such transformed functions. It is known that $f(t)$ lies in $H^p(dt/\pi(1+t^2))$ if and only if $(1+t^2)^{-2/p} f(t)$ lies in $H^p(dt)$ (see [**25**, Chapter 3]).

The *Poisson kernel* $P_z(t)$ for $\mathbf{R}^2_+$ is defined by

$$P_z(t) \;=\; \frac{1}{\pi}\, \frac{v}{(u-t)^2 + v^2}, \qquad z = u + iv,\ v > 0.$$

For $f(t)$ in $H^p(dt/\pi(1+t^2))$, the analytic extension $f(z)$ to $\mathbf{R}^2_+$ is obtained by the convolution

$$f(z) \;=\; f * P_{iv}(u) \;=\; \int_{-\infty}^{\infty} f(t) P_z(t)\, dt,$$

and we sometimes identify $f(t)$ with its extension $f(z)$ in what follows.

A nonzero function in a Hardy space can be factored into a function of modulus one and a function that has no zeros. We restrict our attention to the case of $H^p(dt/\pi(1+t^2))$. An *inner function* is a function in $H^p(dt/\pi(1+t^2))$ that is of modulus one. If w is a nonnegative function in $L^p(dt/\pi(1+t^2))$ such that

$$\int_{-\infty}^{\infty} \log w(t)\, \frac{dt}{1+t^2} \;>\; -\infty\,,$$

and if

$$h(z) \;=\; e^{i\gamma} \exp\left\{ \frac{1}{\pi i} \int_{-\infty}^{\infty} \frac{1+tz}{t-z}\, \log w(t)\, \frac{dt}{1+t^2} \right\}, \qquad z \in \mathbf{R}^2_+\,, \tag{2.1}$$

for some γ in $\mathbf{R}$, then $h(z)$ is called an *outer function* in $H^p(dt/\pi(1+t^2))$. It can be shown that $|h(t)| = w(t)$ on $\mathbf{R}$.

Let z_n ($\neq i$) be a sequence in $\mathbf{R}_+^2$ such that

$$\sum_{n=1}^{\infty} \frac{y_n}{1+|z_n|^2} < \infty, \quad z_n = x_n + iy_n .$$

Then the inner function defined by

$$B(z) = \left(\frac{z-i}{z+i}\right)^m \prod_{n=1}^{\infty} \frac{|z_n^2+1|}{z_n^2+1} \frac{z-z_n}{z-\overline{z_n}}$$

is called a *Blaschke product*. If ν is a positive singular measure on $\mathbf{R}$ such that

$$\int_{-\infty}^{\infty} \frac{1}{1+t^2}\, d\nu(t) < \infty,$$

then the inner function defined by

$$S(z) = \exp\left\{ i\left(\alpha z + \int_{-\infty}^{\infty} \frac{1+tz}{t-z}\, d\nu(t)\right)\right\}$$

is called a *singular inner function*, where $\alpha \geq 0$.

The following canonical factorization theorem is fundamental to the theory of H^p spaces (as discussed in [**8**], [**13**], [**14**], for instance).

THEOREM 2.1. *Every nonzero function $f(z)$ in $H^p(dt/\pi(1+t^2))$ has a unique factorization of the form*

$$f(z) = B(z)S(z)h(z),$$

where $B(z)$ is a Blaschke product, $S(z)$ is a singular inner function, and $h(z)$ is the outer function in $H^p(dt/\pi(1+t^2))$ defined by (2.1) with $w(t) = |f(t)|$. Conversely, every such product $B(z)S(z)h(z)$ belongs to $H^p(dt/\pi(1+t^2))$.

Let A be a commutative Banach algebra with identity 1. Then the class of all maximal ideals of A is called the *maximal ideal space* of A, and denoted by $\mathfrak{M}(A)$. Since there is a one-to-one correspondence between the nonzero complex homomorphisms of A and the maximal ideals in A, it is customary to identify each maximal ideal in $\mathfrak{M}(A)$ with the complex homomorphism that it determines. This allows us to identify $\mathfrak{M}(A)$ with a subset of the unit sphere of the conjugate space A^* of A. We then define the topology of $\mathfrak{M}(A)$ to be the weak-* topology that $\mathfrak{M}(A)$ inherits from A^*, which is the so-called *Gelfand topology* of $\mathfrak{M}(A)$. Then $\mathfrak{M}(A)$ is a compact Hausdorff space. The *Gelfand transform* of f in A is the continuous function $\hat{f}$ on $\mathfrak{M}(A)$ defined by

$$\hat{f}(\xi) = \xi(f), \quad \xi \in \mathfrak{M}(A).$$

Thus the study of certain properties of A reduces to studying the structure of $\mathfrak{M}(A)$.

We see easily that the maximal ideal space $\mathfrak{M}(A(\mathbf{T}))$ of the disc algebra may be considered as the closed unit disc $\overline{D}$. However, when we regard $H^\infty(d\theta/2\pi)$ as a Banach algebra, the structure of $\mathfrak{M}(H^\infty(d\theta/2\pi))$ is very big and complicated. If $z = re^{i\theta}$ lies in D, then each

$$\xi_z(f) = f(re^{i\theta}) = \frac{1}{2\pi}\int_0^{2\pi} f(e^{it})P_r(\theta - t)\, dt, \quad f \in H^\infty(d\theta/2\pi),$$

determines a point in $\mathfrak{M}(H^\infty(d\theta/2\pi))$, where $P_r(\theta)$ is the Poisson kernel for D. This implies that D is naturally embedded as an open subset in $\mathfrak{M}(H^\infty(d\theta/2\pi))$. Carleson's corona theorem [**5**] states that D is dense in $\mathfrak{M}(H^\infty(d\theta/2\pi))$.

For α in $\mathbf{T}$, the *fiber* $\mathfrak{M}_\alpha$ of $\mathfrak{M}(H^\infty(d\theta/2\pi))$ over α is defined to be

$$\mathfrak{M}_\alpha = \{\xi \in \mathfrak{M}(H^\infty(d\theta/2\pi))\,;\xi(z) = \alpha\},$$

where z is the coordinate function. Then we have the decomposition

$$\mathfrak{M}(H^\infty(d\theta/2\pi)) \setminus D = \bigcup_{|\alpha|=1} \mathfrak{M}_\alpha.$$

It is shown that the various fibers $\mathfrak{M}_\alpha$ are homeomorphic to one another. If f lies in $H^\infty(d\theta/2\pi)$, then the *cluster set* of f at α is

$$Cl(f,\alpha) = \bigcap_{r>0} \overline{f(D \cap \Delta(\alpha, r))},$$

where $\Delta(\alpha, r)$ is the open disc with center α and radius r. Then we obtain

$$Cl(f,\alpha) = \hat{f}(\mathfrak{M}_\alpha).$$

Various aspects of H^p theory have been generalized to the abstract setting of uniform algebras, and we give a brief outline. Let Y be a compact Hausdorff space, and let $C(Y)$ be the algebra of all complex-valued continuous functions on Y. We say that a subalgebra A of $C(Y)$ is a *uniform algebra* on Y if A is complete under the uniform norm, contains the constants, and separates the points of Y. A subset E of Y is called a *boundary* for A if

$$\|f\|_\infty = \sup\{|f(x)|\,; x \in E\}, \quad f \in A.$$

There is a smallest closed boundary, which is called the *Šilov boundary* of A. Each complex homomorphism ξ in $\mathfrak{M}(A)$ is represented by a probability measure μ on Y, that is,

$$\hat{f}(\xi) = \int_Y f\,d\mu, \quad f \in A,$$

which is called a *representing measure* for ξ. Then the Hardy space $H^p(\mu), 1 \le p < \infty$, is defined to be the closure of A in $L^p(\mu)$, while the weak-* closure of A in $L^\infty(\mu)$ is $H^\infty(\mu)$. We recall that, in the case of the disc algebra $A(\mathbf{T})$, $H^p(d\theta/2\pi)$ is also obtained in this manner. To use the tools by functional analysis suitably, we need some conditions that assure a certain size of $H^p(\mu)$ in $L^p(\mu)$. So a uniform algebra A is called a *Dirichlet algebra* on Y, if $\operatorname{Re} A = \{\operatorname{Re} f\,; f \in A\}$ is uniformly dense in $C_{\mathbf{R}}(Y)$, the space of all real-valued continuous functions on Y. We say A is a *logmodular algebra* on Y if

$$\log|A^{-1}| = \{\log|f|; f \in A^{-1}\}$$

is uniformly dense in $C_{\mathbf{R}}(Y)$, where A^{-1} denotes the set of invertible elements of A. As is easily seen, every Dirichlet algebra is a logmodular algebra, and $A(\mathbf{T})$ is a Dirichlet algebra on $\mathbf{T}$. However, the uniform algebra of Gelfand transforms of functions in $H^\infty(d\theta/2\pi)$ is not a Dirichlet algebra, but a logmodular algebra on its Šilov boundary.

More generally, let A be a subalgebra of $L^\infty(\mu)$ containing the constants. Then A is called a *weak-* Dirichlet algebra* if μ is multiplicative on A and $A + \overline{A}$ is dense in $L^\infty(\mu)$ in the weak* topology. Many theorems, such as Szegö's theorem and the factorization theorem, have been extended to the context on weak-* Dirichlet algebras. We refer the reader to [**16**], [**17**], [**40**], [**45**] for further details on weak-* Dirichlet algebras.

Let A be a weak-* Dirichlet algebra in $L^\infty(\mu)$. A closed subspace M of $L^p(\mu)$ is *invariant* if AM is contained in M. An invariant subspace M of $L^p(\mu)$ is *simply invariant* if M is not of the form $I_E\, L^p(\mu)$, where I_E is the characteristic function of a set E. We denote by A_0 the subspace of A of all functions f such that

$$\hat{f}(\mu) = \int_Y f\, d\mu = 0.$$

If $A_0 M$ is not dense in M, then M is simply invariant. It is known that, for each h in $H^p(\mu)$, h satisfies *Jensen's inequality*:

$$\log|\hat{h}(\mu)| \le \int_Y \log|h|\, d\mu.$$

When equality holds in finite values, h is called an *outer* function in $H^p(\mu)$. It follows from Szegö's theorem that h is outer if and only if the invariant subspace generated by h equals $H^p(\mu)$, that is, the closure of Ah in $L^p(\mu)$ is $H^p(\mu)$. As before, functions in $H^\infty(\mu)$ of unit modulus are called *inner* functions. It is also known that if f is a function in $H^p(\mu)$ such that $\log|f|$ lies in $L^1(\mu)$, then f can be factored into the form $f = qh$, where q is inner and h is outer. However, factorization theorems have not yet been completely developed, especially for functions f such that $\log|f|$ is not in $L^1(\mu)$.

As before, $\mathfrak{M}(A)$ denotes the maximal ideal space of a uniform algebra A. For ξ and ν in $\mathfrak{M}(A)$, we put

$$\xi \sim \nu \iff \|\xi - \nu\| < 2,$$

where $\|\cdot\|$ is the norm on A^*—in other words, whenever the distance from ξ to ν in A^* is less than 2. Gleason showed that $\sim$ is an equivalence relation on $\mathfrak{M}(A)$. An equivalence class under $\sim$ is called a *Gleason part* of $\mathfrak{M}(A)$. Wermer [**58**] proved that if A is a logmodular algebra, and if a Gleason part P is not a point, then there is a continuous map $\tau(z)$ of the open unit disc D onto P such that $\hat{f}(\tau(z))$ is a bounded analytic function on D for each f in A.

Let $\{z_n\}$ be a sequence in D. We call $\{z_n\}$ an *interpolating* sequence if for each bounded sequence $\{\alpha_n\}$ in $\mathbf{C}$, there is a function f in $H^\infty(d\theta/2\pi)$ such that $f(z_n) = \alpha_n$, $n = 1, 2, \ldots$. By Carleson's characterization [**4**] of interpolating sequences, they play an important role in unraveling the structure of $\mathfrak{M}(H^\infty(d\theta/2\pi))$. A Blaschke product is called an *interpolating Blaschke product* if it has distinct zeros and if these zeros form an interpolating sequence. Using factorization theorems for Blaschke products, Hoffman [**26**] showed that ξ lies in an analytic disc in $\mathfrak{M}(H^\infty(d\theta/2\pi)) \setminus D$ if and only if ξ is in the closure of some interpolating sequence. In view of the size and intractability of $\mathfrak{M}(H^\infty(d\theta/2\pi))$, this would be a great accomplishment.

More information on $\mathfrak{M}(H^\infty(d\theta/2\pi))$ and related results can be found in Izuchi's series of papers [**27**], [**28**], [**29**], [**30**]. In particular, he showed in [**28**] the existence of certain representing measures on a fiber $\mathfrak{M}_\alpha$ under the continuum hypothesis. It would be interesting to investigate cluster sets under this hypothesis.

3. Analytic functions on flows

Let Ω be a compact Hausdorff space on which $\mathbf{R}$ acts as a transformation group. This means that there is a one-parameter group $\{U_t\}_{t\in\mathbf{R}}$ of homeomorphisms on Ω such that the map $(\omega, t) \to U_t\omega$ is continuous on $\Omega \times \mathbf{R}$. The pair $(\Omega, \{U_t\}_{t\in\mathbf{R}})$ is referred to as a (*continuous*) *flow*. For simplicity, the translate $U_t\omega$ of ω by t is denoted by $\omega + t$. A subset M of Ω is *invariant* if M contains $U_t M$ for all t in $\mathbf{R}$. Then Zorn's lemma assures that there is a minimal one in the class of all nonempty closed invariant sets. If Ω itself is minimal, then $(\Omega, \{U_t\}_{t\in\mathbf{R}})$ is called a *minimal flow*. A flow $(\Omega, \{U_t\}_{t\in\mathbf{R}})$ is minimal if and only if the *orbit* of ω,

$$\mathcal{O}(\omega) = \{\omega + t\,; t \in \mathbf{R}\}\,,$$

is dense in Ω for each ω in Ω. On a minimal flow $(\Omega, \{U_t\}_{t\in\mathbf{R}})$, each function ϕ in $C(\Omega)$ is the extension of the uniformly continuous function $t \to \phi(\omega + t)$ on $\mathbf{R}$; so such a function on $\mathbf{R}$ may be regarded as an almost periodic function in a sense.

By the Markov-Kakutani theorem, there is at least one invariant probability measure on Ω. An invariant measure is said to be *ergodic* if every invariant subset of Ω either is negligible or has negligible complement. We use the symbol $\mathcal{K}$ for the set of all invariant probability measures on Ω. Then $\mathcal{K}$ is a nonempty compact convex set in the weak-* topology. Since a measure in $\mathcal{K}$ is ergodic if and only if it is an extreme point of $\mathcal{K}$, the Krein-Milman theorem assures the existence of an ergodic measure in $\mathcal{K}$. A minimal flow $(\Omega, \{U_t\}_{t\in\mathbf{R}})$ is called *strictly ergodic* if there is exactly one invariant probability measure on Ω. A function ϕ in $C(\Omega)$ is *analytic* if $t \to \phi(\omega + t)$ lies in $H^\infty(dt/\pi(1+t^2))$ for each ω in Ω. Let $A(\Omega)$ be the algebra of all analytic functions in $C(\Omega)$. When $(\Omega, \{U_t\}_{t\in\mathbf{R}})$ has no fixed points, $A(\Omega)$ becomes a uniform algebra of Ω, which is called the *uniform algebra induced by* $(\Omega, \{U_t\}_{t\in\mathbf{R}})$. Many results have been obtained about such uniform algebras (see [**9**], [**35**], [**39**], and [**52**]).

Wermer's maximality theorem [**57**] states that the disc algebra $A(\mathbf{T})$ is a maximal closed subalgebra in $C(\mathbf{T})$, that is, every closed subalgebra of $C(\mathbf{T})$ that properly contains $A(\mathbf{T})$ equals $C(\mathbf{T})$. As a generalization of it, Forelli [**11**] proved the following.

THEOREM 3.1. *If* $(\Omega, \{U_t\}_{t\in\mathbf{R}})$ *is minimal, then the induced uniform algebra* $A(\Omega)$ *is a maximal algebra in* $C(\Omega)$.

In connection with this result, he asked the question stated in the Introduction (see [**36**, §6] for a nice account of related topics). In responce to the question, Muhly [**35**] gave, among other things, a sufficient condition for $A(\Omega)$ to be a Dirichlet algebra.

THEOREM 3.2. *If* $(\Omega, \{U_t\}_{t\in\mathbf{R}})$ *is strictly ergodic, then* $A(\Omega)$ *is a Dirichlet algebra on* Ω.

THEOREM 3.3. *Let* μ *be an invariant, ergodic, probability measure on* Ω. *Then* μ *is a representing measure for* $A(\Omega)$ *and* $A(\Omega)$ *is a weak-* Dirichlet algebra in* $L^\infty(\mu)$.

These results are important. Indeed, under these conditions we can discuss Hardy spaces and invariant subspaces by applying the general theory of uniform algebras. It long remained unknown, however, whether $A(\Omega)$ is a Dirichlet algebra for a minimal flow $(\Omega, \{U_t\}_{t\in\mathbf{R}})$. We finally found in [**48**] an example of $A(\Omega)$ that is not a Dirichlet algebra.

On one hand, the minimal flows that are not strictly ergodic have long been investigated in ergodic theory (see, for example, [**42**]). On the other hand, the difference between Dirichlet algebras and logmodular algebras, just like $A(\mathbf{T})$ and $H^\infty(d\theta/2\pi)$, has been discussed. It may be interesting that these two independent areas are related to each other, via the induced uniform algebras.

Let us mention briefly the maximal ideal space $\mathfrak{M}(A(\Omega))$ of $A(\Omega)$. For ϕ in $A(\Omega)$ and $r > 0$, the convolution of ϕ with the Poisson kernel $P_{ir}(t)$,

$$\phi(x,r) = \phi * P_{ir}(x) = \int_{-\infty}^{\infty} \phi(x+t)P_{ir}(t)\,dt, \quad x \in \Omega,$$

is a complex homomorphism of $A(\Omega)$. This shows that $\Omega \times [0,\infty)$ may be considered as a subset of $\mathfrak{M}(A(\Omega))$. As r tends to ∞, $\{(x,r)\}$ accumulates to points in $\mathfrak{M}(A(\Omega))$ for which representing measures are invariant. In particular, when $(\Omega, \{U_t\}_{t\in\mathbf{R}})$ is strictly ergodic, $\mathfrak{M}(A(\Omega))$ is the quotient space obtained from by $\Omega \times [0,\infty]$ by identifying the slice $\Omega \times \{\infty\}$ to a point. Some results relating to the corona theorem have also been obtained. We refer to [**36**], [**47**], and [**59**] for details.

In the rest of this section, we devote ourselves to the construction of an induced uniform algebra that is not a Dirichlet algebra. Let $\beta\mathbf{Z}, S$, and X be as in the Introduction, and let $\{S_t\}_{t\in\mathbf{R}}$ be the one-parameter group of homeomorphisms on X defined by

$$S_t(y,s) = (S^{[t+s]}y, t+s-[t+s]), \quad (y,s) \in X, \tag{3.1}$$

where $[t]$ denotes the largest integer not exceeding t. Then $(X, \{S_t\}_{t\in\mathbf{R}})$ is a flow, which is a special case of the so-called flows built under functions in ergodic theory. Let X_0 be the quotient space obtained from $\mathbf{Z} \times [0,1]$ by identifying $(n,1)$ with $(n+1,0)$. Then X_0 is a dense invariant set in X that is homeomorphic to $\mathbf{R}$. See [**34**] or [**55**] for Stone-Čech compactifications.

Let $C_{ub}(\mathbf{R})$ be the C^*-algebra of all bounded uniformly continuous functions on $\mathbf{R}$. Then we easily obtain the following proposition.

PROPOSITION 3.4. *Let X be the compact Hausdorff space defined above. Then $C_{ub}(\mathbf{R})$ is isometrically isomorphic to $C(X)$. In particular, X is the maximal ideal space of $C_{ub}(\mathbf{R})$.*

Let us consider the uniform algebra $A(X)$ induced by $(X, \{S_t\}_{t\in\mathbf{R}})$. Let f be a function in $H^\infty(dt/\pi(1+t^2))$, and let $r > 0$. Since $f(t+ir)$ is the convolution of $f(t)$ with $P_{ir}(t)$, $t \to f(t+ir)$ lies in $C_{ub}(\mathbf{R})$. By Proposition 3.4, this extends to a function $\phi(x,r)$ in $A(X)$. Conversely, since $\mathbf{R}$ is dense in X, each function in $A(X)$ is determined by its restriction to $\mathbf{R}$. Therefore, $A(X)$ is generated by the family $\{\phi(x,r)\,;r > 0\}$. We next regard $\phi(x,r)$ as a function on $X \times (0,\infty)$. Then a function in $H^\infty(dt/\pi(1+t^2))$ appears on each orbit by Fatou's theorem. Let $H^\infty(X)$ be the space of all boundary functions of $\phi(x,r)$. Then $H^\infty(X)$ is isometrically isomorphic to $H^\infty(dt/\pi(1+t^2))$. We see also that each point in $X \times (0,\infty)$ determines a complex homomorphism for $H^\infty(X)$. Since $H^\infty(dt/\pi(1+t^2))$ and $H^\infty(d\theta/2\pi)$ are isomorphic, $X \times (0,\infty)$ corresponds to a portion of the maximal ideal space $\mathfrak{M}(H^\infty(d\theta/2\pi))$. It is easy to see that $X_0 \times (0,\infty)$ and $X \setminus X_0 \times (0,\infty)$ determine D and the union of the accumulation points of the regions between two circles tangent to $\mathbf{T}$ at 1, respectively.

The maximal ideal space $\mathfrak{M}(A(X))$ contains $X \times [0,\infty)$, which is dense in $\mathfrak{M}(A(X))$ by the corona theorem. Each accumulating point of $X \times [0,\infty)$ is a complex homomorphism of $A(X)$ whose representing measure is invariant. This observation suggests a relation between the corona theorem and the individual ergodic theorem. Indeed, it follows from Wiener's Tauberian theorem that the existence of one side of the following limits implies the existence of the other, and the two are equal:

$$\lim_{r\to\infty} \phi * P_{ir}(x) = \lim_{T\to\infty} \frac{1}{2T}\int_{-T}^{T} \phi(x+t)\,dt.$$

Then the individual ergodic theorem assures the existence of the right side. Recall that each invariant ergodic probability measure is a representing measure for $A(X)$ by Theorem 3.3. With the aid of the ergodic theorem, we may show directly that such representing measures are accumulating points of $X_0 \times (0,\infty)$. Let $\mathcal{K}$ be as above. Then such measures consist of the extreme points of $\mathcal{K}$. Thus if Choquet theory could be developed further, there would be another way to prove the corona theorem.

Let M be a minimal set in $(X, \{S_t\}_{t\in\mathbf{R}})$. Restricting S_t to M, we have a minimal flow $(M, \{S_t\}_{t\in\mathbf{R}})$, which has the strongest topology in the class of all minimal flows. There are consequently a great many invariant probability measures on M.

PROPOSITION 3.5. *Let $(M, \{S_t\}_{t\in\mathbf{R}})$ be as above. Then $(M, \{S_t\}_{t\in\mathbf{R}})$ is universal, meaning that every minimal flow is a factor of $(M, \{S_t\}_{t\in\mathbf{R}})$.*

Outline of Proof. Let $(\Omega, \{U_t\}_{t\in\mathbf{R}})$ be a minimal flow, and fix an ω in Ω. The C^*-algebra generated by $\{t \to \phi(\omega+t)\,;\phi \in C(\Omega)\}$ is a closed subalgebra of $C_{ub}(\mathbf{R})$. It follows from the Banach-Stone theorem that there is a continuous map τ of X onto Ω such that

$$\tau \circ S_t(x) = U_t \circ \tau(x), \quad x \in X.$$

Then the restriction of τ to M is the desired map of M onto Ω. □

THEOREM 3.6. *Let $A(M)$ be the uniform algebra induced by the above flow $(M, \{S_t\}_{t\in\mathbf{R}})$. Then $A(M)$ is a logmodular algebra that is not a Dirichlet algebra.*

Outline of Proof. As is easily seen by the relation to $H^\infty(dt/\pi(1+t^2))$, $A(X)$ is logmodular on X. Since the restriction $A|_M$ of $A(X)$ to M is a subalgebra of $A(M)$, $A(M)$ is also a logmodular algebra on M. We then see that $\mathfrak{M}(A(M))$ is the closure of $M\times[0,\infty)$ in $\mathfrak{M}(A(M))$. To choose a suitable interpolating sequence, we find by Hoffman's theorem a Gleason part P in $M\times\{\infty\}$ that is homeomorphic to D. Let $\overline{P}$ denote the closure of P in $\mathfrak{M}(A(M))$, and let $\hat{A}|_{\overline{P}}$ be the algebra of all restrictions of Gelfand transforms of functions in $A(M)$. Then $\hat{A}|_{\overline{P}}$ is isometrically isomorphic to $H^\infty(d\theta/2\pi)$. As is stated above, $H^\infty(d\theta/2\pi)$ is not a Dirichlet algebra on its Šilov boundary. This implies that there is a nonzero real measure ν on $\overline{P}\setminus P$ that is orthogonal to $\hat{A}|_{\overline{P}}$. Define

$$L(\phi) = \int_{\overline{P}\setminus P} \hat{\phi}(\xi)\,d\nu(\xi), \quad \phi \in C(M), \tag{3.2}$$

where $\hat{\phi}(\xi)$ denotes the integral with respect to the representing measure for ξ. Then L is a bounded linear functional on $C(M)$. It follows from the Riesz-Kakutani theorem that there is an invariant real measure that is orthogonal to $A(M)$. Thus $\mathrm{Re}\,A(M)$ cannot be dense in $C_{\mathbf{R}}(M)$. □

We note that the above ν may be chosen as a measure on the Šilov boundary of $A(M)$, each of which has a representing measure that is invariant and ergodic. It therefore might be interesting to study the relation between (3.2) and the ergodic decomposition of invariant measures. Check Figure 1 again for places of M and P, and [**48**] for the detailed proof of Theorem 3.6 and related topics.

4. Analytic functions on compact groups

In this section we will be concerned with Hardy spaces of analytic almost periodic functions of the form

$$f(t) = \sum_j c_j e^{i\lambda_j t}, \quad t \in \mathbf{R},$$

where the λ_j belong to the positive half of a subgroup Γ of $\mathbf{R}$. If Γ is the group $\mathbf{Z}$ of integers, then f may be regarded as an analytic function on D.

We now assume that Γ is a dense subgroup of $\mathbf{R}$ but endowed with the discrete topology. Let K be the dual group of Γ, and let σ be the normalized Haar measure on K. When λ in Γ is considered as a character on K, it is written χ_λ. A function f in $L^1(\sigma)$ is *analytic* if its Fourier coefficients

$$a_\lambda(f) = \int_K \overline{\chi_\lambda}(x) f(x)\, d\sigma(x)$$

vanish for all negative λ in Γ. The *Hardy space* $H^p(\sigma), 1 \leq p \leq \infty$, is defined to be the space of all analytic functions in $L^p(\sigma)$. Similarly, $A(K)$ denotes the algebra of all analytic functions in $C(K)$. Then $A(K)$ is a Dirichlet algebra on K and σ is a representing measure for $A(K)$. We notice that $H^p(\sigma)$ is also obtained as the closure of $A(K)$ in $L^p(\sigma)$ as in Section 2.

For each t in $\mathbf{R}$, e_t is the element of K defined by $e_t(\lambda) = e^{i\lambda t}$ for any λ in Γ. Then the map $t \to e_t$ embeds $\mathbf{R}$ continuously onto a dense subgroup of K. We then define the one-parameter group $\{T_t\}_{t\in\mathbf{R}}$ of homeomorphisms on K by

$$T_t x = x + e_t, \quad x \in K.$$

The flow $(K, \{T_t\}_{t\in\mathbf{R}})$ is strictly ergodic, and σ is the unique invariant probability measure on it. Since

$$\chi_\lambda(T_t x) = \chi_\lambda(x + e_t) = e^{i\lambda t}\chi_\lambda(x),$$

$\chi_\lambda(x)$ is an eigenfunction with eigenvalue λ. We remark that the discrete spectrum theorem characterizes ergodic flows that are conjugate to $(K, \{T_t\}_{t\in\mathbf{R}})$ (see [**15**], [**56**]).

Fix a positive γ in Γ, and let K_γ be the subgroup consisting of all x in K such that $\chi_\gamma(x) = 1$. Then K may be identified with $K_\gamma \times [0, 2\pi/\gamma)$ via the map $y+s \to (y, s)$. When 2π lies in Γ, we fix a homeomorphism on $K_{2\pi}$ by $Sy = y + e_1$. Then $\{T_t\}_{t\in\mathbf{R}}$ may be represented in the form (3.1).

We see easily that a function f in $L^p(\sigma)$ is analytic if and only if $t \to f(x+t)$ lies in $H^p(dt/\pi(1+t^2))$, for a.e. x in K. Using this fact, we characterize outer functions in $H^2(\sigma)$ as follows.

LEMMA 4.1. *A function h in $H^2(\sigma)$ is outer if and only if $\hat{h}(\sigma) \neq 0$ and $t \to h(x+t)$ is outer in $H^2(dt/\pi(1+t^2))$ for a.e. x in K.*

Outline of Proof. Suppose that $A(K)h$ is not dense in $H^2(\sigma)$. Then there is a nonconstant function g in $H^2(\sigma)$ that is orthogonal to $A(K)h$. Since $t \to \overline{g(x+t)}h(x+t)$ lies in $H^1(dt/\pi(1+t^2))$, $t \to h(x+t)$ cannot be an outer function in $H^2(dt/\pi(1+t^2))$ for a.e. x in K. Conversely, suppose that the set of all x in K such that

$$\log|h * P_{ir}(x)| < \int_{-\infty}^{\infty} \log|h(x+t)|\, P_{ir}(t)\, dt$$

has positive measure. Since $h * P_{ir}(x)$ lies in $H^2(\sigma)$, we obtain

$$\log|\int_K h * P_{ir}(x) d\sigma(x)| \leq \int_K \log|h * P_{ir}(x)|\, d\sigma(x)$$

by Jensen's inequality. This implies that

$$\log|\hat{h}(\sigma)| < \int_K \log|h(x)|\, d\sigma(x).$$

Thus h cannot be outer in $H^2(\sigma)$. □

Let f be a function in $C(K)$ with Fourier series

$$f(x) \sim \sum_n a_n \chi_{\lambda_n}(x).$$

Then f is uniformly approximated by a sequence of trigonometric polynomials on X by the Stone-Weierstrass theorem. This implies that $F_x(t) = f(x+t)$ is a uniformly almost periodic function (in the sense of Bohr) with exponents in Γ,

$$F_x(t) = \sum_n a_n \chi_{\lambda_n}(x) e^{i\lambda_n t}.$$

If $y + e_{t_j}$ tends to x for a fixed y in K, then $F_x(t)$ may be regarded as a kind of limit of translations $F_y(t+t_j)$. The same relation holds between $L^2(\sigma)$-functions and almost periodic functions in the sense of Besicovitch, if we exclude an exceptional null set.

Since the 1950s, $A(K)$ and $H^p(\sigma)$ have been explored as extensions of $A(\mathbf{T})$ and $H^p(d\theta/2\pi)$ to functions of several variables. In particular, the invariant subspace theory has flourished in this setting. Each invariant subspace in $L^2(\sigma)$ corresponds to a function of modulus one on $K \times \mathbf{R}$, which is called a *cocycle*. However, there remain many interesting problems; for example, we do not know whether every simply invariant subspace can be generated by one of its elements.

Let $\mathbf{T}^\omega$ be the complete direct sum of countably many copies of $\mathbf{T}$. If Γ is countable, then K is a closed subgroup of $\mathbf{T}^\omega$ (see [**44**] for details). We now adduce an example that plays an important role in the next section.

EXAMPLE. Let $\mathbf{Z}^\infty$ be the direct sum of countably many copies of $\mathbf{Z}$, that is, the group of all sequences of which all but a finite number of terms vanish. Taking primes p as indices, we denote $\mathbf{Z}^\infty$ by

$$\mathbf{Z}^\infty = \mathbf{Z}_2 \oplus \mathbf{Z}_3 \oplus \mathbf{Z}_5 \oplus \cdots \oplus \mathbf{Z}_p \oplus \cdots,$$

where $\mathbf{Z}_p = \mathbf{Z}$. Then $\mathbf{Z}^\infty$ has a natural order, and its dual group is identified with $\mathbf{T}^\omega$. Indeed, let τ be the isomorphism of $\mathbf{Z}^\infty$ into $\mathbf{R}$ by

$$\tau(\{n_p\}) = \sum_{p\,;\,prime} n_p \log p, \quad \{n_p\} \in \mathbf{Z}^\infty.$$

We see that $\tau(\mathbf{Z}^\infty)$ is the dense subgroup $\Gamma = \{\log r\,; r \text{ is positive rational}\}$ of $\mathbf{R}$, and that the dual group K of Γ is isomorphic to $\mathbf{T}^\omega$ via the map τ^* defined by

$$\langle \tau^*(x), \{n_p\}\rangle = \langle x, \tau(\{n_p\})\rangle.$$

We also let

$$\mathbf{T}^{\omega} = \mathbf{T}_2 \otimes \mathbf{T}_3 \otimes \mathbf{T}_5 \otimes \cdots \otimes \mathbf{T}_p \otimes \cdots,$$

where $\mathbf{T}_p = \mathbf{T}$. Since $e_t(\log p) = e^{it\log p}$, the one-parameter group $\{T_t\}_{t\in\mathbf{R}}$ of homeomorphisms on $\mathbf{T}^{\omega}$ is defined by

$$T_t(\{e^{i\theta_p}\}) = \{e^{i(\theta_p + t\log p)}\}, \quad \{e^{i\theta_p}\} \in \mathbf{T}^{\omega}.$$

We note that the normalized Haar measure σ_P on $\mathbf{T}^{\omega}$ is given by

$$d\sigma_P = \prod_{p;\ prime} \frac{1}{2\pi}\, d\theta_p\,.$$

We identify K with $\mathbf{T}^{\omega}$ as usual.

Analyticity on compact groups with ordered duals is developed in [**13**, Chapter VII], [**22**], and [**44**, Chapter 8].

5. Extension of the Riemann zeta-function and Dirichlet series with Euler products

By a *Dirichlet series*, we mean a series of the form

$$f(s) = \sum_{n=1}^{\infty} \frac{a_n}{n^s}, \quad s = \sigma + it, \tag{5.1}$$

where $\{a_n\}$ is a sequence in $\mathbf{C}$. Then the series converges on a right half-plane $\sigma > \sigma_0$, on which the sum $f(s)$ of the series is an analytic function. The most important Dirichlet series is the *Riemann zeta-function*, which is the case when $a_n = 1$:

$$\zeta(s) = \sum_{n=1}^{\infty} \frac{1}{n^s} = \prod_{p;\ prime} (1 - \frac{1}{p^s})^{-1}, \quad \sigma > 1\,.$$

This infinite product is called the *Euler product*. The function $\zeta(s)$ can be analytically continued over $\mathbf{C}$ except when $s = 1$, where there is a simple pole with residue 1. The *Riemann Hypothesis* amounts to the assertion that $\zeta(s)$ has no zeros in the half-plane $\sigma > 1/2$. If we assume that $\{a_n\}$ is bounded and $a_{mn} = a_m a_n$ in (5.1), then $f(s)$ is also representable as an Euler product:

$$f(s) = \prod_{p;\ prime} (1 - \frac{a_p}{p^s})^{-1}, \quad \sigma > 1. \tag{5.2}$$

In particular, when $|a_n| = 1$, (5.2) is deeply connected with $\zeta(s)$. Indeed, such an $f(s)$ may be obtained as some kind of limit of translations of $\zeta(s)$ in a vague sense.

Using the flow $(\mathbf{T}^{\omega}, \{T_t\}_{t\in\mathbf{R}})$ in the above example, we first extend $\zeta(s)$ to an analytic function on $\mathbf{T}^{\omega}$. Let $u > 1/2$. Since

$$\zeta(u + it) = \sum_{n=1}^{\infty} \frac{1}{n^u}\, e^{-it\log n},$$

$t \to \zeta(u - it)$ can be considered as an analytic almost periodic function on $\mathbf{R}$. So this may be extended to an analytic function on $\mathbf{T}^{\omega}$ by

$$Z_u(x) = \sum_{n=1}^{\infty} \frac{1}{n^u}\, \chi_{\log n}(x), \quad x \in \mathbf{T}^{\omega}\,. \tag{5.3}$$

Since

$$\|Z_u\|_2^2 = \sum_{n=1}^{\infty} (\frac{1}{n^u})^2 < \infty,$$

$Z_u(x)$ lies in the Hardy space $H^2(\sigma_P)$ defined in Section 4. Restricting $Z_u(x)$ to the orbit $\mathcal{O}(0) = \{e_t\,; t \in \mathbf{R}\}$ and extending it to $\{\operatorname{Im} z > 1/2\}$, we have $\zeta(s)$, $\sigma > 1$. To state things more precisely, we denote by $Z_u(x+z) = Z_u(x+e_z)$ the analytic extension of $t \to Z_u(x+e_t)$ to $\mathbf{R}_+^2$. Then we obtain

$$\begin{aligned} Z_u(e_z) &= \sum_{n=1}^{\infty} \frac{1}{n^u} \chi_{\log n}(e_z) \\ &= \sum_{n=1}^{\infty} \frac{1}{n^u} e^{i(\log n)z} \\ &= \sum_{n=1}^{\infty} \frac{1}{n^u} \frac{1}{n^{-iz}}\,. \end{aligned}$$

If we put $z = is$ (the 90° rotation of the right half-plane), then $Z_u(e_z)$ represents $\zeta(s+u)$. Thus $Z_u(x)$ is regarded as an extension of $\zeta(s)$ to $\mathbf{T}^\omega$, and each $z \to Z_u(x+z), \operatorname{Im} z > 0$, is analogous to $\zeta(s)$ by its almost periodicity. In particular, if $\operatorname{Im} z > 1/2$, then $Z_u(x+z)$ is continuous on $\mathbf{T}^\omega$ as a function of x; so $\zeta(s)$ is determined completely by $z \to Z_u(x+z)$ (see Figure 2).

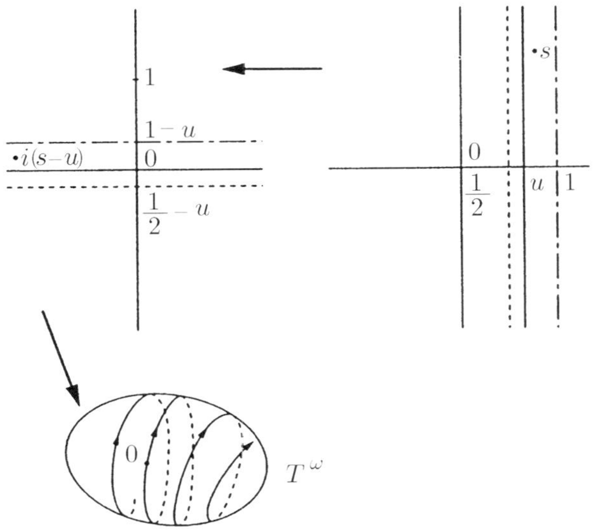

FIGURE 2

LEMMA 5.1. *Let $u > 1/2$, and let $Z_u(x)$ be the function in (5.3). Then both $Z_u(x)$ and $Z_u(x)^{-1}$ are outer functions in $H^2(\sigma_P)$.*

Outline of Proof. Recall that the Möbius function $\mu(n)$ is defined by

$$\mu(n) = \begin{cases} 1, & \text{if } n=1, \\ (-1)^n, & \text{if } n \text{ is the product of } k \text{ different primes}, \\ 0, & \text{otherwise.} \end{cases}$$

Then

$$\frac{1}{\zeta(s)} = \sum_{n=1}^{\infty} \frac{\mu(n)}{n^s}, \quad \sigma > 1. \tag{5.4}$$

If we put

$$Z_u(x)^{-1} = \sum_{n=1}^{\infty} \frac{\mu(n)}{n^u} \chi_{\log n}(x), \quad x \in \mathbf{T}^{\omega},$$

then $Z_u(x)^{-1}$ lies in $H^2(\sigma_P)$ and $Z_u(x)\,Z_u(x)^{-1} = 1$. From these facts, it follows that both subspaces $H^{\infty}(\sigma_P)\,Z_u(x)$ and $H^{\infty}(\sigma_P)\,Z_u(x)^{-1}$ are dense in $H^2(\sigma_P)$. □

By Lemmas 4.1 and 5.1, $z \to Z_u(x+z)$ is an outer function in $H^2(dt/\pi(1+t^2))$ for a.e. x in $\mathbf{T}^{\omega}$; so it has no zeros. The restriction of $Z_u(x+z)$ to the orbit $\mathcal{O}(0)$ of 0 represents $\zeta(s)$ as mentioned above. Since $\zeta(s)$ has the pole $s = 1$, $\mathcal{O}(0)$ is contained wholly in the exceptional null set in Lemma 4.1.

Let us look into a property of $Z_u(x+z)$ precisely. Since

$$\begin{aligned} Z_u(x+z) &= \sum_{n=1}^{\infty} \frac{1}{n^u} \cdot \chi_{\log n}(x+e_z) \\ &= \sum_{n=1}^{\infty} \frac{1}{n^u} \cdot \chi_{\log n}(x)\, e^{i(\log n)z} \\ &= \sum_{n=1}^{\infty} \frac{a_n}{n^u} \frac{1}{n^{-iz}}, \end{aligned}$$

where we put $a_n = \chi_{\log n}(x)$, it follows that the sequence $\{a_n\}$ satisfies $|a_n| = 1$ and $a_m \cdot a_n = a_{mn}$ by the definition of characters. If $n = p_1^{b_1} \cdot p_2^{b_2} \cdots p_l^{b_l}$ is the factorization of n in prime numbers, then we obtain

$$a_n = a_{p_1}^{b_1} \cdot a_{p_2}^{b_2} \cdots a_{p_l}^{b_l}. \tag{5.5}$$

Conversely, each $x = \{a_p\}$ in $\mathbf{T}^{\omega}$ decides $\chi_{\log n}(x)$ by (5.5). Thus we have the following.

THEOREM 5.2. *Define a_n by (5.5) with a given $\{a_p\}$ in $\mathbf{T}^{\omega}$, and let*

$$f(s) = \sum_{n=1}^{\infty} \frac{a_n}{n^s} = \prod_{p;\ prime} \left(1 - \frac{a_p}{p^s}\right)^{-1}, \quad s = \sigma + it, \tag{5.6}$$

which is absolutely convergent on $\sigma > 1$. Then, for a.e. $\{a_p\}$ in $\mathbf{T}^{\omega}$, $f(s)$ extends analytically to $\sigma > 1/2$ and has no zeros.

In a similar way, we see that $Z_u(x)^{-1}$ has the same property. However, when $\{a_p\}$ in $\mathbf{T}^{\omega}$ is given concretely, we have no idea for deciding whether $f(s)$ in (5.6) satisfies the conclusion of Theorem 5.2.

Let $x_0 = \{-1\}$ in $\mathbf{T}^\omega$. Since $\chi_{\log p}(x_0) = -1$, we see that

$$\frac{\zeta(s)}{\zeta(2s)} = \prod_{p;\ prime} (1 - \frac{\chi_{\log p}(x_0)}{p^s}), \quad s = \sigma + it\,.$$

This implies that

$$Z_u(x)\, Z_{2u}(2x)^{-1} = Z_u(x + x_0)^{-1}.$$

Since $Z_{2u}(2x)^{-1}$ is represented as an absolutely convergent Fourier series, $Z_{2u}(2x)^{-1}$ is a continuous outer function on $\mathbf{T}^\omega$. So $Z_u(x)$ acts as if it were self-reciprocal by the translation $x \to x + x_0$. The set of zeros of $Z_u(x + z)$ corresponds to the set of poles of $Z_u(x + x_0 + z)$, although both are empty for a.e. x, by Theorem 5.2. However, using a property of alternating series, we can easily find Dirichlet series of the forms (5.6) that have zeros, poles, or both in $1/2 < \sigma \leq 1$.

The Rademacher functions seem to be useful. Indeed, in a way similar to [**8**, Appendix A], it can be shown that, for a.e. $\{a_p\}$ in $\mathbf{T}^\omega$, the Dirichlet series $f(s)$ of the form (5.6) has the property that the limit

$$\lim_{\sigma \to \frac{1}{2}} f(\sigma + it)$$

does not exist for dt-a.e. t. Thus, unlike the case of $\zeta(s)$, such Dirichlet series cannot satisfy any kind of functional equation and cannot even extend analytically across the line $\sigma = 1/2$. Moreover, a property of the Rademacher functions enables us to show that both $Z_u(x)$ and $Z_u(x)^{-1}$ lie in $H^p(\sigma_P), 1 \leq p < \infty$ (see [**20**] for another proof).

There is a relation between the abscissa of convergence for Dirichlet series and Hardy spaces on $\mathbf{R}$. Let $f(s)$ be a Dirichlet series of the form (5.1). It follows from [**50**, 9.14] that if the function

$$F(\lambda) = \sum_{\log n \leq \lambda} a_n$$

satisfies $F(\log n) = O(n^u)$, then $f(s)$ converges for $\sigma > u$.

THEOREM 5.3. *Let $f(s)$ be a Dirichlet series of the form* (5.1). *Suppose that $f(s)$ extends analytically to $\sigma > \sigma_0\,(\geq 0)$ and*

$$t \to f(\sigma_0 - it) \in H^r(dt/\pi(1 + t^2))$$

for $r > 2$. Then $f(s)$ converges for $\sigma > \sigma_0 + 1/r$.

Outline of Proof. For each q with $2 \leq q < r$, we choose p such that $1/q + 1/p = 1$. Since

$$|\sigma + it|^p = |\sigma + it|^{\frac{2p}{r}} \cdot |\sigma + it|^{p(1 - \frac{2}{r})},$$

we obtain by Hölder's inequality that

$$\int_{-\infty}^{\infty} \left|\frac{f(\sigma + it)}{\sigma + it}\right|^p dt \leq \left\{\int_{-\infty}^{\infty} \frac{|f(\sigma + it)|^r}{|\sigma + it|^2}\, dt\right\}^{p/r} \left\{\int_{-\infty}^{\infty} |\sigma + it|^{\frac{-p(r-2)}{r-p}}\, dt\right\}^{\frac{r-p}{r}}.$$

Since $r > q = p/(p-1)$, we see that $p(r-2)/(r-p) > 1$. This assures that the last integral converges. On the other hand, we may see that the Fourier transform of $\lambda \to e^{-\lambda\sigma} F(\lambda)$ is

$$\frac{f(\sigma + it)}{\sigma + it}.$$

Then it follows from the Young-Hausdorff theorem that

$$\int_{-\infty}^{\infty} e^{-\lambda\sigma q} \, |F(\lambda)|^q \, d\lambda \; < \; \infty.$$

This implies that

$$\sum_{n=1}^{\infty} e^{-\sigma q \log(n+1)} \, |F(\log n)|^q \, \{\log(n+1) - \log n\} \; < \; \infty \, .$$

Since the terms of this series tend to 0, we have

$$|F(\log n)| \; = \; O(n^{\sigma + 1/q}) \, ;$$

so $f(s)$ has the desired property. □

As we mentioned above, $Z_u(x)$ lies in $H^p(\sigma_P), 1 \le p < \infty$. Then $t \to Z_u(x+t)$ lies in $H^p(dt/\pi(1+t^2))$. We thus obtain the following.

PROPOSITION 5.4. *Under the hypotheses of Theorem 5.2, the Dirichlet series of the form* (5.6) *converges on* $\sigma > 1/2$, *for a.e.* $\{a_p\}$ *in* $\mathbf{T}^\omega$.

We know little about the relation between the exceptional null sets in Theorem 5.2 and Proposition 5.4. Littlewood's theorem [**51**, 12.25] states that the convergence of (5.4) is equivalent to the Riemann Hypothesis. Regarding this, it would be interesting to investigate such null sets in further detail.

Finally, we would like to make a remark on mean values of negative powers of $\zeta(s)$. Let $0 < p < \infty$ and let $u > 1/2$. Since $Z_u(x)^{-1}$ is outer by Proposition 5.1, we see easily that $Z_u(x)^{-p}$ is defined and lies in $H^1(\sigma_P)$. It then follows from the individual ergodic theorem that

$$\lim_{T\to\infty} \frac{1}{T} \int_0^T \, |Z_u(x+t)|^{-p} \, dt \; = \; \int_{\mathbf{T}^\omega} |Z_u(x)|^{-p} \, d\sigma_P(x)$$

holds for a.e. x in $\mathbf{T}^\omega$. Although we do not know whether this holds on the orbit $\mathcal{O}(0)$, it is possible to show a weak version with the aid of a normal family argument and Rouché's theorem. A subset J of the nonnegative integers $\mathbf{Z}^+$ has *density zero* if

$$\frac{1}{N} \, (J \cap \{\, 0, 1, 2, \cdots, N-1 \,\}) \to 0 \quad (\text{as } N \to \infty).$$

We fix an $\ell > 0$. Then there is a subset J of $\mathbf{Z}^+$ of density zero such that

$$\lim_{N\to\infty} \frac{1}{N\ell} \sum_{\substack{n=0 \\ J \not\ni n}}^{N-1} \int_{n\ell}^{(n+1)\ell} \, |\zeta(u+it)|^{-p} \, dt \; = \; \int_{\mathbf{T}^\omega} |Z_u(x)|^{-p} \, d\sigma_P(x).$$

Of course, we can derive the counterpart for the case of positive powers of $\zeta(s)$. For the proof, see our subsequent note, *Dirichlet series induced by the Riemann zeta-function.*

In [**1**], [**3**], and [**41**], the Riemann zeta-function is discussed in connection with functional analysis. In Helson's series of notes [**18**], [**19**], [**20**], [**21**], one can find a modern treatment of Dirichlet series by the methods of functional analysis.

The author would like to express his sincere gratitude to Professor Koji Matsumoto for his useful comments. He was kind enough to inform us that methods similar to those of Section 5 are used in the value distribution theory of the Riemann zeta-function, and also to introduce Laurinčikas' book [**33**] as a good reference source in the field. He also gave another proof of Theorem 5.2. Thanks are due to the referee as well, for valuable suggestions that improved the first version of this paper.

References

[1] H. Bercovici and C. Foias, *A real variable restatement of Riemann's hypothesis,* Israel J. Math. **48** (1984), 57-68.

[2] A. Beurling, *On two problems concerning linear transformations in Hilbert space,* Acta Math. **81** (1949), 239-255.

[3] A. Beurling, *A closure problem related to the Riemann Zeta-function,* Proc. Nat. Acad. Sci. U.S.A. **41** (1955), 312-314.

[4] L. Carleson, *An interpolation problem for bounded analytic functions,* Amer. J. Math. **80** (1958), 921-930.

[5] L. Carleson, *Interpolations by bounded analytic functions and the corona problem,* Ann. of Math. **76** (1962), 547-559.

[6] M. Denker and E. Eberlein, *Ergodic flows are strictly ergodic,* Advances in Math. **13** (1974), 437-473.

[7] N. Dunford and J. Schwartz, *Linear Operators, Part I,* Wiley Interscience, New York, 1958.

[8] P. Duren, *Theory of H^p Spaces,* Academic Press, New York and London, 1970.

[9] F. Forelli, *Analytic and quasi-invariant measures,* Acta Math. **118** (1967), 33-59.

[10] F. Forelli, *Fourier theory and flows,* Actes Congrès Internat. Math. (Nice, 1970), vol. 2, Gauthier-Villars, Paris, 1971, pp. 475-476.

[11] F. Forelli, *A maximal algebra,* Math. Scand. **30** (1970), 152-158.

[12] T. Gamelin, *H^p spaces and extremal functions in H^1,* Trans. Amer. Math. Soc. **124** (1966), 158-157.

[13] T. Gamelin, *Uniform Algebras,* Prentice-Hall, Englewood Cliffs, N.J., 1969.

[14] J. Garnett, *Bounded Analytic Functions,* Academic Press, New York, 1981.

[15] P. Halmos, *Lectures on Ergodic Theory,* Mathematical Society of Japan, Tokyo, 1956.

[16] M. Hasumi, *On shift-invariant subspaces,* Sūgaku **17** (1966), 22-32 (in Japanese).

[17] M. Hasumi, *Theory of invariant subspaces,* Sūgaku **28** (1976), 47-59 (in Japanese).

[18] H. Helson, *Convergent Dirichlet series,* Ark. Mat. **4** (1962), 501-510.

[19] H. Helson, *Foundations of the theory of Dirichlet series,* Acta Math. **118** (1967), 61-77.

[20] H. Helson, *Compact groups and Dirichlet series,* Ark. Mat. **8** (1969), 139-143.

[21] H. Helson, *Convergence of Dirichlet series,* Lecture Notes in Mathematics, 336, Springer-Verlag, Berlin, 1973, pp. 153-160.

[22] H. Helson, *Analyticity on compact abelian groups,* in *Algebras in Analysis,* Academic Press, London, 1975, pp. 1-62.

[23] H. Helson, *Compact groups with ordered duals VII,* J. London Math. Soc. (2) **20** (1979), 509-515.

[24] H. Helson and J. Tanaka, *Singular cocycles and the generator problem,* Operator Theoretical Methods (Timişoara, 1998; A. Gheondea et al., eds.), Theta Foundation, Bucharest, 2000, pp. 173-186.

[25] K. Hoffman, *Banach Spaces of Analytic Functions,* Prentice-Hall, Englewood Cliffs, N.J., 1962.

[26] K. Hoffman, *Bounded analytic functions and Gleason parts,* Ann. of Math. **86** (1967), 74-111.

[27] K. Izuchi, *Zero sets of interpolating Blaschke products,* Pacific J. of Math. **119** (1985), 337-342.

[28] K. Izuchi, *QC-level sets and questions of Douglas algebras.* J. Funct. Anal. **65** (1986), 293-308.

[29] K. Izuchi, *Interpolating sequences in a homeomorphic part of H^∞,* Proc. Amer. Math. Soc. **111** (1991), 1057-1065.

[30] K. Izuchi, *Spreading Blaschke products and homeomorhic parts,* in Harmonic/Analytic Function Spaces and Linear Operators, Surikaisekikenkyusho-Kokyuroku No. 1049 (1998), 157–161 (in Japanese). MR **99i:**00018

[31] T. Kano *et al, Riemann Hypothesis,* Nihon-Hyōronsya, 1991 (in Japanese).

[32] N. Kurokawa,*A quarter of a millennium since Euler products*, Gendai-Sūgaku no Ayumi 4, 1992, 13-27 (in Japanese).

[33] A. Laurinčikas, *Limit Theorems for the Riemann Zeta-Function,* Mathematics and its Applications **352** Kluwer Academic Publishers Group, Dordrecht, 1996.
[34] K. Morita, *Theory of Topological Spaces*, Iwanami-Zensyo, Tokyo, 1981 (in Japanese).
[35] P. Muhly, *Function algebras and flows,* Acta Sci. Math. (Szeged) **35** (1973), 111-121.
[36] P. Muhly, *Function algebras and flows II,* Ark. Mat. **11** (1973), 203-213.
[37] P. Muhly, *Function algebras and flows IV,* Trans. Amer. Math. Soc. **203** (1975), 55-66.
[38] P. Muhly, *Isometries of ergodic Hardy spaces,* Israel J. Math. **36** (1980), 50-74.
[39] P. Muhly, *The function-algebraic ramifications of Wiener's work on prediction theory and random analysis,* in Norbert Wiener, *Collected Works with Commentaries,* Vol. III, MIT Press, Cambridge, MA, 1981, pp. 339-370.
[40] T. Nakazi, *Invariant subspaces of weak-* Dirichlet algebras,* Pacific J. Math. **69** (1977), 151-167.
[41] N. Nikolski, *Distance formulae and invariant subspaces, with an application to localization of zeros of the Riemann ζ-function,* Ann. Inst. Fourier, Grenoble **45** (1995), 143-151.
[42] J. Oxtoby, *Ergodic sets,* Bull. Amer. Math. Soc. **58** (1952), 116-136.
[43] W. Parry, *Compact abelian group extensions of discrete dynamical systems,* Z. Wahrscheinlichkeitstheorie Verw. Geb. **13** (1969), 95-113.
[44] W. Rudin, *Fourier Analysis on Groups,* Interscience, New York, 1962.
[45] T. Srinivasan and J. Wang, *Weak-* Dirichlet algebras,* in *Function Algebras,* Scott-Foresman, Chicago, 1966, pp. 216-249.
[46] O. Takenouchi, *On Banach algebras,* Sūgaku **28** (1976), 1-9 (in Japanese).
[47] J. Tanaka, *Corona problem and flows,* J. Funct. Anal. **102** (1991), 360-378.
[48] J. Tanaka, *Flows in fibers,* Trans. Amer. Math. Soc. **343** (1994), 779-804.
[49] J. Tanaka, *Single generator problem,* Trans. Amer. Math. Soc. **348** (1996), 4113-4129.
[50] E.C. Titchmarsh, *The Theory of Functions,* 2nd ed., Oxford University Press, 1939.
[51] E.C. Titchmarsh, *The Theory of the Riemann Zeta-Function* (revised by D. R. Heath-Brown), Oxford University Press, 1988.
[52] J. Tomiyama, *Function algebras and flows,* Sūgaku **28** (1976), 35-46 (in Japanese).
[53] J. Wada, *Normed Rings*, Kyōritsu-Shyuppan, Tokyo, 1969 (in Japanese).
[54] J. Wada, *Some fundamental concepts of function algebras*, Sūgaku **28** (1976), 9-17 (in Japanese).
[55] R. Walker, *The Stone-Čech Compactification,* Springer-Verlag, Berlin, 1974.
[56] P. Walters, *An Introduction to Ergodic Theory,* Springer-Verlag, New York, 1982.
[57] J. Wermer, *On algebras of continuous functions,* Proc. Amer. Math. Soc. **4** (1953), 866-869.
[58] J. Wermer, *Dirichlet algebras,* Duke Math. J. **27** (1960), 373-382.
[59] J. Xia, *Conditional expectations and the corona problem of ergodic Hardy spaces*, J. Funct. Anal. **64** (1985), 251-274.

Department of Mathematics, School of Education, Waseda University, Shinjuku, Tokyo 169-8050, Japan
E-mail address: `jtanaka@mn.waseda.ac.jp`

Translated by JUN-ICHI TANAKA

Amer. Math. Soc. Transl.
(2) Vol. **204**, 2001

Painlevé Equations in the Past 100 Years

Hiroshi Umemura

1. The discovery of the Painlevé equations

> Les Mathématiques constituent un continent solidement agencé, dont tous les pays sont bien reliés les uns aux autres; l'œuvre de Paul Painlevé est une île originale et splendide dans l'ocean voisin.
>
> H. Poincaré ([H], p. 51)

What impression would the reader receive from this remark? We can admit here a certain criticism but Poincaré (1854–1912) was a tried friend of Painlevé (1863–1933).

As explained in [O2] and [U2], the motivation of the discovery of the Painlevé equations was a pursuit of special functions. At the end of the 19th century, there was a trend of searching for new special functions through the study of algebraic ODEs $y'' = R(t, y, y')$ without movable singularities, where t is the independent variable, $y' = dy/dt$, $y'' = d^2y/dt^2$, and $R(t, y, y')$ is a rational function in t, y, y'. They studied the algebraic differential equation $y'' = R(t, y, y')$ of the second order because their aim was to generalize the algebraic differential equation

$$y'^2 = 4y^3 - g_2 y - g_3,$$

which is satisfied by the Weierstrass $\wp$-function.

The basic idea depends on an optimistic viewpoint that if the order of the differential equation increases as $2, 3, \ldots$, we would encounter special functions one after another. Painlevé first listed all such ODEs $y'' = R(t, y, y')$, and then erased those whose solution can be written with commonly known functions. This is surely a natural process, in accordance with the objective of finding new special functions. Consequently, he obtained the six equations P_{I}, P_{II}, $\ldots$, P_{VI}, known today as the Painlevé equations (see (17.1)). This is actually a tough job, requiring much more calculation than it seems to.

Working over the Painlevé equations for over a decade, we can conclude it is truely a miracle that the significant mathematical objects were discovered by Painlevé's method. Poincaré's remark quoted above appears, in spite of all his understanding of the difficulty of the problem solved by Painlevé, to imply his

---

This article originally appeared in Japanese in Sūgaku **51** (4) (1999), 395–420.
2000 *Mathematics Subject Classification.* Primary 34M55; Secondary 32Gxx, 33E17, 37Ixx.

inevitable doubt on its value. In a word, he wants to ask the following serious question.

PROBLEM 1.1. Do the equations P_{I}, P_{II}, ... , P_{VI} really define any special functions?

It is not guaranteed that the Painlevé equations discovered by the above procedure necessarily define special functions.

Another fundamental question is: what are special functions? A function deserves to be called a special function only if it has a certain profundity and an inexhaustible bewitching power, just as the theta functions, the most typical special functions, do for instance.

The meaning of problem 1.1 becomes clear if we investigate third-order algebraic differential equations $F(t, y, y', y'', y^{(3)}) = 0$ without movable singularities. There are works of Chazy, Garnier, Bureau, etc., but classification remains yet to be done. Even worse, there arises a fundamental question, that is, whether imposing the condition that the equation has no movable singularity is adequate or not. In fact, as the work of Jacobi shows, the theta constants $\theta_2(\tau)$, $\theta_3(\tau)$ and $\theta_4(\tau)$ satisfy a third-order algebraic differential equation $G(\tau, y, y', y'', y^{(3)}) = 0$. This equation seems quite natural and hence significant, but it has movable singularities (see [J], [C1]).

We have worked on problem 1.1 for a long time. The conclusion is that the Painlevé equations do define special functions. It was only two or three years ago that we were convinced, when relations were revealed between the Painlevé equations and combinatorics of the Young diagrams or representation theory of the general linear groups (see sections 13 and 14).

Some 100 years have passed since the discovery of the Painlevé equations. Developments during this century at last allow us to answer problem 1.1 affirmatively. I am sure every mathematician studying the Painlevé equations would be eager to show Painlevé and Poincaré the achievements that have been made in this century.

2. Irreducibility of the Painlevé equations

The Painlevé equations were obtained through a refinement process of throwing away, from the list of algebraic differential equations $y'' = R(t, y, y')$ without movable singularities, those whose solutions are written by commonly known functions (henceforth referred to as classical functions). Since the discovery of the equations, therefore, it was expected that their solutions cannot generally be represented with classical functions. Namely, they are irreducible to classical functions.

As for the first equation, soon after the discovery of the equations, in 1902, Painlevé announced its irreducibility. A dispute took place between R. Liouville and Painlevé about this announcement. Disputes are unusual in the history of mathematics. Painlevé terminated it (see section 7 of [U2] for the details) with a suggestion of using an infinite-dimensional differential Galois theory proposed by Drach.

It was not until 1988 that the irreducibility of P_{I} was proved ([N], [U1], [U2]). The equations P_{II}, P_{III}, ... , P_{VI} contain parameters, and for some of the values the solutions of the equations can be written with classical functions. It is therefore an important problem to determine the classical solutions of the equations (see [U3]).

More precisely P_{VI}, ... , P_{II} contain the hypergeometric, the confluent hypergeometric, the Hermite, the Bessel and the Airy functions, respectively, as their classical solutions. The Painlevé equations therefore involve many orthogonal polynomials, such as the Jacobi polynomials, the Legendre polynomials, the Laguerre polynomials, the Hermite polynomials, etc. As shown in sections 8–12, the Painlevé equations generate non-classical special polynomials too.

3. Isomonodromic deformations

The first remarkable result in the history of the Painlevé equations was obtained by R. Fuchs in 1905. He discovered that the sixth equation P_{VI} describes an isomonodromic deformation of a linear ODE. In other words, P_{VI} is derived from an isomonodromic deformation of a linear ODE. This shows that P_{VI} has another origin other than the classification of algebraic ODEs $y'' = R(t, y, y')$ without movable singularities, which we saw in section 1. This discovery gave a smashing blow to Poincaré's doubt expressed in his remark quoted at the beginning of this article.

The Painlevé equations were born in the traditional French mathematics of the 19th century, of which dullness was the target of a violent criticism of the young generation led by A. Weil after WWI, but it turned out that they are closely related with the prominent tradition of Gauss and Riemann at Göttingen. As a consequence, Poincaré's doubt crumbled partly and it became apparent that Painlevé's work is not "an isolated island in the neighboring ocean."

His work was taken over by R. Garnier, who showed that P_{I}, P_{II}, ... , P_{VI} all describe isomonodromic deformations. This should be an important contribution forming a connection between the analysis of France and the tradition of Göttingen. Bourbaki seems, however, to have regarded his work as an out-of-date mathematics. Garnier actually treated some topics that suit Bourbaki's taste, such as dynamics on the Jacobians of spectral curves. But once a confrontation arises between groups of mathematicians, it seems difficult to look open-mindedly at works of the other group.

After the era of discovery in the early 20th century, the Painlevé equations disappeared from the forefront of mathematics. The underground water passage of isomonodromic deformation theory appeared again in physics in the 1970s, opening a new epoch of the Painlevé equations.

4. Hamiltonian representations of the Painlevé equations

Each Painlevé equation has a Hamiltonian representation. For example, for the second equation $P_{\mathrm{II}}(\alpha)\colon y'' = 2y^3 + ty + \alpha$, where α is a parameter, its Hamiltonian representation is as follows. Let $H_{\mathrm{II}}(b, t, q, p) = p^2/2 - \left(q^2 + t/2\right) p - bq$, where $b = \alpha + 1/2$, and consider a Hamiltonian system

$$\frac{dq}{dt} = \frac{\partial H_{\mathrm{II}}}{\partial p}, \quad \frac{dp}{dt} = -\frac{\partial H_{\mathrm{II}}}{\partial q},$$

i.e.,

$$S_{\mathrm{II}}(b) \begin{cases} dq/dt = p - q^2 - t/2, \\ dp/dt = 2pq + b. \end{cases}$$

We see that $P_{\mathrm{II}}(\alpha)$ and $S_{\mathrm{II}}(b)$ are equivalent. Namely, if y is a solution of $P_{\mathrm{II}}(\alpha)$, then $(q, p) = (y, y' + y^2 + t/2)$ is a solution of $S_{\mathrm{II}}(b)$, and if (q, p) is a solution of $S_{\mathrm{II}}(b)$, then $y = q$ is a solution of $P_{\mathrm{II}}(\alpha)$.

The other Painlevé equations also have similar Hamiltonian representations. This fact had been known since long ago to the Belarussian school. It was Kazuo

Okamoto who made the first systematic study of them. The following problem arises naturally.

PROBLEM 4.1. Where does the Hamiltonian structure of the Painlevé equations come from?

Some progress has recently been made on that problem (see [MMT], [ST], and [SU]). We can show the existence and uniqueness of the Hamiltonian structure through algebro-geometric study of the space of initial conditions but we do not go into details.

5. Symmetries in the Painlevé equations

We take once again the system $S_{\mathrm{II}}(b)$ equivalent to the second Painlevé equation as an example. We denote the group of affine transformations of the line $\mathbb{C}$ by Aff $\mathbb{C}$. Let us consider the following affine transformations of $\mathbb{C}$:

$$t_+(b) = b+1, \quad t_-(b) = b-1, \quad i(b) = -b \qquad (b \in \mathbb{C}).$$

Let

$$G := \langle t_+, t_-, i \rangle \subset \operatorname{Aff}\mathbb{C},$$

i.e., G is a subgroup of Aff $\mathbb{C}$, generated by t_+, t_- and i. It is clear that $\langle t_+, t_- \rangle \simeq \mathbb{Z}$, $\langle i \rangle \simeq \mathbb{Z}/2\mathbb{Z}$ and $A = \langle i \rangle \ltimes \langle t_+, t_- \rangle \simeq S_2 \ltimes \mathbb{Z}^2$. That is, G is the Weyl group$^{(1)}$ of the affine root system of type $\tilde{A}_1$. Since $G \subset \operatorname{Aff}\mathbb{C}$, G acts on the line $\mathbb{C} = \{b \mid b \in \mathbb{C}\}$, the parameter space of $S_{\mathrm{II}}(b)$. We can lift this operation to transformations of solutions of the system S_{II} of differential equations.

Namely, let $Sol_{\mathrm{II}}(b)$ be the set of all solutions of $S_{\mathrm{II}}(b)$, and set $Sol_{\mathrm{II}} := \coprod_{b\in\mathbb{C}} Sol_{\mathrm{II}}(b)$, where $\coprod$ denotes disjoint union. For $(q,p) \in Sol_{\mathrm{II}}(b)$, set

$$I(b,-b)(q,p) := \left(q + \frac{b}{p}, p\right);$$

$$T_+(b, b+1)(q,p) := \left(-q - \frac{b}{p}, -p + 2\left(q + \frac{b}{p}\right)^2\right);$$

$$T_-(b, b-1)(q,p) := \left(-q + \frac{b-1}{p - 2q^2 - t}, -p + 2q^2 + t\right).$$

Then we have

$$I(b,-b)(q,p) \in Sol_{\mathrm{II}}(-b),$$

$$T_+(b,b+1)(q,p) \in Sol_{\mathrm{II}}(b+1),$$

$$T_-(b,b-1)(q,p) \in Sol_{\mathrm{II}}(b-1),$$

and we thereby get mappings

$$I(b,-b)\colon\ Sol_{\mathrm{II}}(b) \to Sol_{\mathrm{II}}(-b),$$

$$T_+(b,b+1)\colon\ Sol_{\mathrm{II}}(b) \to Sol_{\mathrm{II}}(b+1),$$

$$T_-(b,b-1)\colon\ Sol_{\mathrm{II}}(b) \to Sol_{\mathrm{II}}(b-1).$$

Rigorously speaking, we should pay attention to the case where the denominators vanish. For example, if $p = 0$ in the definition of $I(b,-b)(q,p)$, then we can show $b = 0$. Hence the rigorous definition of $I(b,-b)(q,p)$ must be

$$I(b,-b)(q,p) = \begin{cases} (q + b/p, p) & \text{for } b \neq 0, \\ (q,p) & \text{for } b = 0. \end{cases}$$

We define a mapping $I\colon\ Sol_{\mathrm{II}} \to Sol_{\mathrm{II}}$ from $I(b,-b)\colon\ Sol_{\mathrm{II}}(b) \to Sol_{\mathrm{II}}(-b)$ in the following manner. For $(q,p) \in Sol_{\mathrm{II}}$, there exists a unique number b such that $(q,p) \in Sol_{\mathrm{II}}(b)$, and $I(b,-b)$ is defined by

$$I(q,p) = I(b,-b)(q,p) \in Sol_{\mathrm{II}}(-b) \subset Sol_{\mathrm{II}}.$$

Similarly, we define $T_+\colon\ Sol_{\mathrm{II}} \to Sol_{\mathrm{II}}$ and $T_-\colon\ Sol_{\mathrm{II}} \to Sol_{\mathrm{II}}$ so that we have

$$T_- \circ T_+ = T_+ \circ T_- = Id_{Sol_{\mathrm{II}}}, \quad I \circ T_+ = T_- \circ I, \quad I^2 = Id_{Sol_{\mathrm{II}}}.$$

It follows from these identities that I, T_+ and T_- determine isomorphisms $Sol_{\mathrm{II}} \to Sol_{\mathrm{II}}$, and that $\tilde{G}$, a subgroup of $Aut(Sol_{\mathrm{II}})$ generated by these isomorphisms, is isomorphic to the affine Weyl group G, where $Aut(Sol_{\mathrm{II}})$ is the group of automorphisms of the set Sol_{II}.

This implies that the affine Weyl group G acts on the set Sol_{II}, and that if a map $\pi\colon\ Sol_{\mathrm{II}} \to \mathbb{C}$ is defined by $\pi(q,p) = b \in \mathbb{C}$ for $(q,p) \in Sol_{\mathrm{II}}(b) \subset Sol_{\mathrm{II}}$, then the projection π is equivariant under the actions of G. The transformation on solutions of $S_{\mathrm{II}}(b)$, induced by an element of the affine Weyl group, is called a Bäcklund transformation.

6. Yablonskii-Vorob′ev polynomials

The system $S_{\mathrm{II}}(b)$ of differential equations has a unique rational solution when $b \in 1/2 + \mathbb{Z}$ ([U3], Theorem 6.1). For $b = 1/2$, $(q,p) = (0, t/2)$ is the rational solution of $S_{\mathrm{II}}(1/2)$. By iterating T_+ or T_- on this solution, we can obtain the rational solution of $S_{\mathrm{II}}(n + 1/2)$ for $n \in \mathbb{Z}$. For example,

$$T_-\left(\frac{1}{2}, -\frac{1}{2}\right)\left(0, \frac{t}{2}\right) = \left(\frac{1}{t}, \frac{t}{2}\right) \in Sol_{\mathrm{II}}\left(-\frac{1}{2}\right),$$

$$T_-\left(-\frac{1}{2}, -\frac{3}{2}\right)\left(\frac{1}{t}, \frac{t}{2}\right) = \left(-\frac{1}{t} + \frac{3t^2}{t^3+4}, \frac{t^3+4}{2t^2}\right) \in Sol_{\mathrm{II}}\left(-\frac{3}{2}\right).$$

So far as efficiency of calculation is concerned, this method is not nice. Yablonskii and Vorob′ev invented a much more elegant way. Consider the recurrence formula

(6.1)
$$T_n(t) = \frac{1}{T_{n-2}(t)}\left[tT_{n-1}^2 - 4\left(T''_{n-1}(t)T_{n-1}(t) - T'_{n-1}(t)^2\right)\right] \qquad \text{for } n \geq 2$$

with initial conditions $T_0(t) = T_1(t) = 1$. Recurrence formula (6.1) determines rational functions $T_n(t)$ for $n \geq 0$. Namely, $T_0(t) = T_1(t) = 1$, $T_2(t) = t$, $T_3(t) = 4 + t^3$, $T_4(t) = -80 + 20t^3 + t^6$, and so forth ([U3]). We have the following

PROPOSITION 6.2. *The rational function $T_n(t)$ is a monic polynomial with integral coefficients.*

It is an interesting problem to give a purely algebraic proof of this proposition ([FOU], [T1], [KO1]), because when we generate polynomials by the other Painlevé equations, we have to show that the rational function defined by a recurrence formula is a polynomial.

Yablonskii and Vorob′ev showed that for $n \geq 0$, $T'_{n+1}(t)/T_{n+1}(t) - T'_n(t)/T_n(t)$ is the rational solution of $P_{\mathrm{II}}(-n)$: $y'' = 2y^3 + ty - n$. In terms of the system $S_{\mathrm{II}}(b)$, this shows that if we set

$$q := T'_{n+1}/T_{n+1} - T'_n/T_n, \quad p := q' + q^2 + t/2,$$

then (q, p) is a solution of $S_{\mathrm{II}}(1/2 - n)$. This method is more refined than the direct iteration of the transformation $T_-(b, b-1) : Sol_{\mathrm{II}}(b) \to Sol_{\mathrm{II}}(b-1)$.

By solving equation (6.1) with respect to $T_{n-2}(t)$, we can calculate $T_{n-2}(t)$ if we know $T_n(t)$ and $T_{n-1}(t)$. Therefore, setting $T_0(t) = T_1(t) = 1$, we can thereby determine the entire series of $T_n(t)$, which turn out to be polynomials and satisfy $T_{n+1}(t) = T_{-n}(t)$. Moreover, for an arbitrary integer n, we can show that $T'_{n+1}/T_{n+1} - T'_n/T_n$ is the rational solution of $P_{\mathrm{II}}(-n)$.

7. τ-functions

Let $(q(t), p(t)) \in Sol_{\mathrm{II}}(b)$. Namely, $(q(t), p(t))$ is a solution of the system $Sol_{\mathrm{II}}(b)$. We define a τ-function associated with the solution $(q(t), p(t))$ by

$$\tau'(t)/\tau(t) = H_{\mathrm{II}}(b, t, q(t), p(t)),$$

where $H_{\mathrm{II}}(b, t, q, p) = p^2/2 - \left(q^2 + t/2\right)p - bq$ is the Hamiltonian (section 4, $S_{\mathrm{II}}(b)$).

Now suppose that a solution $(q_v, p_v) \in Sol_{\mathrm{II}}(v)$ is given for $v \in \mathbb{C}$. We denote the solution obtained by applying the Bäcklund transformation T_- m times to the solution (q_v, p_v) by $(p_{v-m}, q_{v-m}) \in Sol_{\mathrm{II}}(v-m)$. Namely, we set $(q_{v-m}, p_{v-m}) = T_-^m(q_v, p_v)$.

Let τ_m be the τ-function associated with the solution (q_{v-m}, p_{v-m}). Then, by definition, $\tau'_m/\tau_m = H(v-m, t, q_{v-m}, p_{v-m})$.

THEOREM 7.1. *$\{\tau_m\}_{m\in\mathbb{Z}}$ satisfies the Toda equation; i.e., for an arbitrary integer m,*

$$\frac{d^2}{dt^2}\log\tau_m = c(m)\frac{\tau_{m_1}\tau_{m+1}}{\tau_m^2}, \tag{7.1.1}$$

where $c(m) \in \mathbb{C}^*$ ([OJ], $J = \mathrm{III}$).

Let us see an example of the theorem.

EXAMPLE 7.2. For $v = 1/2$, it follows from $(q_{1/2}, p_{1/2}) = (0, t/2) \in Sol_{\mathrm{II}}(1/2)$ and $H(b, t, q, p) = p^2/2 - \left(q^2 - t/2\right) - bq$, that

$$H(1/2, t, q_{1/2}, p_{1/2}) = -t^2/8.$$

Since $\tau'_0/\tau_0 = -t^2/8$, we have

$$\tau_0 = const.\exp\left[\int(-t^2/8)dt\right] = const.\exp\left[-t^3/24\right]. \tag{7.3}$$

On the other hand,

$$(q_{-1/2}, p_{-1/2}) = T_-(1/2, -1/2)(0, t/2) = (1/t, t/2) \in Sol_{\mathrm{II}}(-1/2).$$

By a simple calculation, we see that $H(-1/2, t, q_{-1/2}, p_{-1/2}) = -t^2/8$. By the same method as for τ_0, we have

$$\tau_1 = const. \exp\left[-t^3/24\right]. \tag{7.4}$$

We set

$$\tau_m(t) = \exp\left[-t^3/24\right] T_m(t), \tag{7.5}$$

and we may assume $T_0(t) = T_1(t) = 1$ by virtue of (7.3) and (7.4). Moreover, we assume $c(m) = 1/4$ and substituting (7.5) into the Toda equation, we obtain the recurrence formula

$$T_{m+1} = \frac{1}{T_{m-1}}\left[tT_m^2 - 4\left(T_m'' T_m - (T_m')^2\right)\right], \tag{7.6}$$

which is nothing but the recurrence formula (6.1).

On the other hand, generally for $(q_b, p_b) \in Sol_{\mathrm{II}}\,(b)$, let us set

$$(q_{b-1}, p_{b-1}) := T_-(b, b-1)(q_b, p_b) \in Sol_{\mathrm{II}}\,(b-1).$$

We see again, by a simple calculation, that

$$H(b-1, t, q_{b-1}, p_{b-1}) = H(b, t, q_b, p_b) + q_b. \tag{7.7}$$

Setting $b = 1/2 - m$ and applying (7.7) to $(q_{1/2-m}, p_{1/2-m})$ above, we get

$$\begin{aligned} &H(1/2-m-1, t, q_{1/2-m-1}, p_{1/2-m-1}) \\ &\quad = H(1/2-m, t, q_{1/2-m}, p_{1/2-m} + q_{1/2-m}) + q_{1/2-m}. \end{aligned}$$

Therefore, by the definition of the τ-function, we have that

$$\tau_{m+1}'/\tau_{m+1} = \tau_m'/\tau_m + q_{1/2-m},$$

i.e.,

$$q_{1/2-m} = \tau_{m+1}'/\tau_{m+1} - \tau_m'/\tau_m.$$

By (7.5), we have

$$q_{1/2-m} = T_{m+1}'/T_{m+1} - T_m'/T_m.$$

This is the reason why we can calculate the rational solutions of $P_{\mathrm{II}}(-m)$ by using the Yablonskii-Vorob′ev polynomials. As we see below, since this method allows us to generate special polynomials from a solution of the other Painlevé equations, we sum up the above procedure.

(7.6.1). Start with the rational solution $(q, p) = (0, t/2) \in Sol_{\mathrm{II}}\,(b)$ with $b = 1/2$.

(7.6.2). Iterate the translation $b \mapsto b - 1$ on the initial solution $(q, p) = (0, t/2)$ to obtain a series of rational solutions of $S_{\mathrm{II}}(b)$.

(7.6.3). The series of the τ-functions obtained through the iteration of translations generally satisfy the Toda equation.

(7.6.4). Define the functions $T_m(t)$ by the relation $\tau_m(t) = \exp\left[-t^3/24\right] T_m(t)$, where $\{\tau_m\}$ is the series of the τ-functions obtained from the series of rational solutions in (7.6.2), and substitute these τ_m's into the Toda equation. Then the $T_m(t)$'s satisfy recurrence formula (7.6).

We notice further that the parameter values $b \in 1/2 + \mathbb{Z}$, for which we have rational solutions and for which we constructed the Yablonskii-Vorob′ev polynomials, are the fixed points of an element of the affine Weyl group of type $\tilde{A}_1$. For example, $b = 1/2$ is the fixed point of the affine transformation

$$i \circ t_- : \mathbb{C} \to \mathbb{C}, \quad v \mapsto 1 - v,$$

and the corresponding Bäcklund transformation

$$I(v-1, 1-v) \circ T_-(v, v-1)\colon \mathit{Sol}_{\mathrm{II}}(v) \to \mathit{Sol}_{\mathrm{II}}(v), \quad (q,p) \mapsto (-q, -p + 2q^2 + t)$$

induces

$$I(-1/2, 1/2) \circ T_-(1/2, -1/2)\colon \mathit{Sol}_{\mathrm{II}}(1/2) \to \mathit{Sol}_{\mathrm{II}}(1/2), \tag{7.8}$$

which leaves the solution $(0, t/2) \in \mathit{Sol}_{\mathrm{II}}(1/2)$ invariant. This holds for every rational solution of $\mathit{Sol}_{\mathrm{II}}(b)$ with $b \in 1/2 + \mathbb{Z}$. This suggests that if we try to generate interesting polynomials from a Painlevé equation, first we look for a parameter fixed by an element g of the affine Weyl group and then a solution left invariant by the Bäcklund transformation associated with g.

Now, before we generate special polynomials from the Painlevé equations, we give the general definition of the τ-functions. The Jth Painlevé equation P_J ($\mathrm{I} \le J \le \mathrm{VI}$) is equivalent to a Hamiltonian system

$$S_J(\mathbf{v}) \begin{cases} dq/dt = \partial H_J/\partial p, \\ dp/dt = -\partial H_J/\partial q, \end{cases}$$

where the Hamiltonian $H_J(\mathbf{v}, t, q, p)$ is a polynomial in $\mathbf{v}$, q, p, and is a rational function in t, and $\mathbf{v}$ is a vector-valued parameter (cf. [U3], section 3). For a given solution $(q(t), p(t))$ of $S_J(\mathbf{v})$, we define the τ-function associated with $(q(t), p(t))$ by $\tau'(t)/\tau(t) = H_J(\mathbf{v}, t, q(t), p(t))$.

8. Okamoto polynomials

In just the same way as the rational solutions of P_{II} generate the Yablonskii-Vorob′ev polynomials, those of P_{IV} generate the Okamoto polynomials.

P_{IV} is equivalent to a Hamiltonian system

$$S_{\mathrm{IV}}(v_1, v_2, v_3) \begin{cases} dq/dt = \partial H_{\mathrm{IV}}/\partial p, \\ dp/dt = -\partial H_{\mathrm{IV}}/\partial q \end{cases}$$

with Hamiltonian

$$H_{\mathrm{IV}}(v_1, v_2, v_3) = 2qp^2 - \left(q^2 + 2tq - 2(v_1 - v_2)\right)p - (v_1 - v_3)q - (2qp + 2v_1)t,$$

where

$$(v_1, v_2, v_3) \in V_{\mathrm{IV}} = \left\{(v_1, v_2, v_3) \in \mathbb{C}^3 \mid v_1 + v_2 + v_3 = 0\right\}$$

is a parameter (see [U3]).

The extended affine Weyl group G of type $\tilde{A}_2$ acts on the parameter space V_{IV}. We can lift the action of G on the parameter space V_{IV} to an action of G on the solution space of $S_{\mathrm{IV}}(v_1, v_2, v_3)$ as the Bäcklund transformations, just as in P_{II}. Consider a transformation in G,

$$g\colon V_{\mathrm{IV}} \to V_{\mathrm{IV}}, \quad (v_1, v_2, v_3) \mapsto (v_3 + 2/3, v_1 - 1/3, v_2 - 1/3).$$

This transformation has a unique fixed point $\frac{1}{3}(1,0,-1) \in V_{\mathrm{IV}}$, and g induces a transformation of solutions,

$$Sol_{\mathrm{IV}}(v_1, v_2, v_3) \to Sol_{\mathrm{IV}}(v_3 + 2/3, v_1 - 1/3, v_2 - 1/3), \tag{8.1}$$

where $Sol_{\mathrm{IV}}(\mathbf{v})$ denotes the set of all solutions of $S_{\mathrm{IV}}(\mathbf{v})$. So, in particular, (8.1) defines a transformation

$$Sol_{\mathrm{IV}}\, \tfrac{1}{3}(1,0,-1) \to Sol_{\mathrm{IV}}\, \tfrac{1}{3}(1,0,-1). \tag{8.2}$$

If we try to find a solution fixed by the transformation (8.2), we get a rational solution

$$(q(t), p(t)) = (-2t/3, t/3 + 1/(2t)) \in Sol_{\mathrm{IV}}\, \tfrac{1}{3}(1,0,-1).$$

We start with this rational solution and proceed according to the procedure shown in (7.6).

Now we consider the series of solutions $(q_n(t), p_n(t))$, which we obtain by iterative application of the translation $V_{\mathrm{IV}} \to V_{\mathrm{IV}}$, $\mathbf{v} \mapsto \mathbf{v} + n/3\,(-1,-1,2)$ to the rational solution above. Let τ_n be the τ-function associated with $(q_n(t), p_n(t))$.

Here is the Toda equation for the fourth Painlevé equation P_{IV}.

THEOREM 8.3. *For every integer n, we have the Toda equation*

$$\frac{d^2}{dt^2} \log \tau_n + 2\,(n - 2/3) = c(n) \frac{\tau_{n-1}\tau_{n+1}}{\tau_n^2}, \tag{8.3.1}$$

where $c(n) \in \mathbb{C}^*$ ([O III], Proposition 3.5).

Now, let $\tau_n(t) = \exp\left(t^4/27 - (n-1)t^2/3\right) T_n(t)$ and let $c(n) = 1$ for all integers $n \geq 0$. Then we get a recurrence formula

$$T_{n+1}(t) = \frac{1}{T_{n-1}(t)} \left[T_n''(t)T_n(t) - T_n'(t)^2 + \left(4t^2/9 + 2(2n-1)/3\right) T_n(t)^2\right].$$

By a simple calculation we see that $T_0(t) = T_1(t) = 1$.

Letting now $x = \sqrt{3/2}\, t$, $S_n(x) = T_n\left(\sqrt{2/3}\, x\right)$ and $c(n) = 2/3$ for all n, we get a recurrence formula

$$S_{n+1}(x) = \frac{1}{S_{n-1}(x)} \left[S_n''(x)S_n(x) - S_n'(x)^2 + \left(x^2 + (2n-1)\right) S_n(x)^2\right]$$

with initial condition $S_0(x) = S_1(x) = 1$. The $S_n(x)$ are called Okamoto polynomials (see (17.2)), and we can show that they are monic polynomials for every integer $n \geq 0$.

9. Special polynomials generated by algebraic solutions of P_{III}

The third Painlevé equation P_{III} is equivalent to a Hamiltonian system

$$S_{\mathrm{III}}(v_1, v_2) \begin{cases} t\, dq/dt = 2q^2p - q^2 - v_1 q + t, \\ t\, dp/dt = -2qp^2 + 2qp - v_1 p - (v_1 + v_2)/2, \end{cases}$$

with Hamiltonian

$$H_{\mathrm{III}}(\mathbf{v}, t, q, p) = \frac{1}{t}\left(q^2p^2 - (q^2 + v_1 q - t)p + \frac{(v_1 + v_2)q}{2}\right),$$

where $\mathbf{v} = (v_1, v_2) \in \mathbb{C}$ is a parameter. Let $\mathrm{Aff}\,\mathbb{C}^2$ be the group of affine transformations of the plane $\mathbb{C}^2$, and let i be the reflection of $\mathbb{C}^2$ with respect to the line $v_1 + v_2 = 0$, and j the reflection with respect to the line $v_1 - v_2 = 0$. Let

$t_{(1,1)}$ be the translation of $\mathbb{C}^2$ by the vector $(1,1)$, and $t_{(-1,1)}$ the translation by the vector $(-1,1)$. The group G of Bäcklund transformations of the system $S_{\mathrm{III}}(v_1,v_2)$ is a subgroup $\langle i,j,t_{(1,1)},t_{(-1,1)}\rangle \subset \operatorname{Aff}\mathbb{C}^2$. From the matrix representation of G in terms of the basis $\{u_1=(1,1), u_2=(1,-1)\}$, we see that

$$G = (\text{affine Weyl group of type } \tilde{A}_1) \times (\text{affine Weyl group of type } \tilde{A}_1).$$

The composition $i \circ t_{(1,1)} : \mathbb{C}^2 \to \mathbb{C}^2$ of affine transformations is the reflection of $\mathbb{C}^2$ with respect to the line $v_1+v_2+1=0$, and when the parameter (v_1,v_2) is on this line, we seek the solutions of $S_{\mathrm{III}}(v_1,v_2)$ fixed by the corresponding Bäcklund transformation to get

$$(q(t),p(t)) = \left(s, 1+\frac{1+2v_1}{4s}\right), \tag{9.1}$$

where $v_1+v_2+1=0$ and $s^2=t$ so that (9.1) is an algebraic solution of $S_{\mathrm{III}}(v_1,v_2)$ with $v_1+v_2+1=0$. Starting from the solution (9.1), applying the Bäcklund transformation associated with the transformation $\mathbb{C}^2\to\mathbb{C}^2$, $\mathbf{v}\mapsto\mathbf{v}+n(1,1)$, we get a solution of $S_{\mathrm{III}}(\mathbf{v})$ yielding a series of the τ-functions $\{\tau_n\}_{n\in\mathbb{Z}}$.

Its Toda equation is

$$\frac{d}{dt}\left[t\frac{d}{dt}\log\tau_n\right] = c(n)\frac{\tau_{n-1}\tau_{n+1}}{\tau_n^2},$$

where $c(n)\in\mathbb{C}^*$ (see [O IV], p. 328, formula (3.12)).

Let

$$\tau_n(t) = \exp\left[\int\left(H_{\mathrm{III}}(\mathbf{v},t,q(t),p(t)) - \frac{n}{4}(2v_1+1)t^{-1}\right)dt\right]T_n(t)$$

$(v_1+v_2+1=0)$. Substituting this in the Toda equation with the assumption $c(n)=1$ for all n, we get a recurrence formula

$$T_{n+1}(s) = \frac{1}{4sT_{n_1}}\left[-\left(\frac{dT_n}{ds}+s\frac{d^2T_n}{ds^2}\right)T_n + s\left(\frac{dT_n}{ds}\right)^2 + (4s+2v_1+1)T_n^2\right]$$

with initial condition $T_0(s)=T_1(s)=1$.

Since $T_n(s)$ in the equation above is not a polynomial in s, we let $u=1/(4s)$, $a=2v_1+1$ and $S_n(a,u)=T_n(v,s)$ so that the recurrence formula becomes

$$S_{n+1}(a,u) = \frac{1}{S_{n-1}}\left[-\left(4u^3\frac{\partial S_n}{\partial u}+4u^4\frac{\partial^2 S_n}{\partial u^2}\right)+4u^4\left(\frac{\partial S_n}{\partial u}\right)^2+(1+au)S_n^2\right]$$

with initial condition $S_0(a,u)=S_1(a,u)=1$. We can also prove that $S_n(a,u)$ is a polynomial in two variables a and u (see (17.3)).

10. Special polynomials associated with a rational solution of P_{V}

Let

$$\begin{aligned}A &= q(q-1)^2p^2 + \left((v_1-v_2)(q-1)^2 + 2(v_1+v_2)q(q-1) - tq\right)p\\ &\quad + (v_1-v_3)(v_1-v_4)(q-1).\end{aligned}$$

The fifth Painlevé equation P_{V} is equivalent to a Hamiltonian system

$$S_{\mathrm{V}}(\mathbf{v})\begin{cases} dq/dt = \partial H_{\mathrm{V}}/\partial p,\\ dp/dt = -\partial H_{\mathrm{V}}/\partial p,\end{cases}$$

with Hamiltonian $H_{\rm V} = A/t$, where the parameter $\mathbf{v} = (v_1, v_2, v_3, v_4)$ is an element of $V_{\rm V} = \left\{ \mathbf{v} = (v_1, v_2, v_3, v_4) \in \mathbb{C}^4 \,\middle|\, \sum_{i=1}^4 v_i = 0 \right\}$. The Bäcklund transformation group G of the system $S_{\rm V}(\mathbf{v})$ is the extended affine Weyl group of type $\tilde{A}_3$.

An affine transformation

$$f\colon V_{\rm V} \to V_{\rm V},\ (v_1, v_2, v_3, v_4) \mapsto (v_4 + 3/4, v_1 - 1/4, v_2 - 1/4, v_3 - 1/4)$$

is an element of order 4 of G, and has a unique fixed point $\frac{1}{8}(3, 1, -1, -3)$. On the other hand, the set of fixed points of f^2 coincides with a line

$$L = \{(v, 1/2 - v, v - 1/2, -v) \in V_{\rm V} \mid v \in \mathbb{C}\}.$$

Since L contains a parameter, we look for those solutions fixed by f^2 that will generate special polynomials involving a parameter and will accordingly be more interesting.

We can find the fixed solution $(q(t), p(t)) = (-1, t/8 + v)$ on the line L. Mapping it by the Bäcklund transformations associated with the translations $V_{\rm V} \to V_{\rm V}$, $\mathbf{v} \mapsto \frac{n}{4}(1, -3, 1, 1)$, we get a series of solutions $\{(q_n(t), p_n(t))\}_{n\in\mathbb{Z}}$, and therefore a series of the τ-functions $\{\tau_n\}_{n\in\mathbb{Z}}$. The Toda equation satisfied by the series of the τ-functions of the system $S_{\rm V}(\mathbf{v})$ of differential equations is

$$\frac{d}{dt} t \left[\frac{d}{dt} \log \tau_n \right] + v_4 - v_1 + n = c(n) \frac{\tau_{n-1}\tau_{n+1}}{\tau_n^2},$$

where $c(n) \in \mathbb{C}^*$ ([U2], p. 68, formula (2.9)).

Let

$$\tau_n(t) = \exp\left[\int \left\{ \frac{t}{16} + v + \frac{2v(2v-1)}{t} + \frac{n}{t}\left(\frac{t}{4} - 2v - 1 \right) \right\} dt \right] T_n(t)$$

and $c(n) = 1$ for all n. Substituting $\tau_n(t)$ into the Toda equation above, we get a recurrence formula

$$T_{n+1}(v, t) = \frac{1}{T_{n-1}} \left[\left(\frac{t}{8} - v + \frac{3n}{4} \right) T_n^2 + \left(\frac{\partial T_n}{\partial t} \right) T_n + t \left(\frac{\partial^2 T_n}{\partial t^2} \right) T_n - t \left(\frac{\partial T_n}{\partial t} \right)^2 \right].$$

A calculation shows that $T_0(v, t) = T_1(v, t) = 1$ and that every $T_n(v, t)$ is a polynomial of two variables v and t (see (17.4)).

11. Special polynomials associated with algebraic solutions of $P_{\rm VI}$

The sixth Painlevé equation $P_{\rm VI}$ is equivalent to a Hamiltonian system

$$S_{\rm VI}(\mathbf{b}) \begin{cases} dq/dt = \partial H_{\rm VI}/\partial p, \\ dp/dt = -\partial H_{\rm VI}/\partial p, \end{cases}$$

with Hamiltonian

$$\begin{aligned} H_{\rm VI} = \frac{1}{t(t-1)} \big[& q(q-1)(q-t)p^2 \\ & - \{(b_1 + b_2)(q-1)(q-t) + (b_1 - b_2)q(q-t) + (b_3 + b_4)q(q-1)\} p \\ & + (b_1 + b_3)(b_1 + b_4) \big], \end{aligned}$$

where $\mathbf{b} = (b_1, b_2, b_3, b_4) \in \mathbb{C}^4$ is a parameter. Its Bäcklund transformation group G is the extended affine Weyl group of type $\tilde{D}_4$ ([O1]).

A transformation $g\colon \mathbb{C}^4 \to \mathbb{C}^4$, $\mathbf{b} \mapsto (b_1, b_2, -b_3 - 1, -b_4)$ belongs to G, and the set of its fixed points is a plane $\{(b_1, b_2, -1/2, 0) \mid b_1, b_2 \in \mathbb{C}\}$. Seeking the solution $(q(t),\ p(t))$ fixed by the Bäcklund transformation associated with g on this plane, we get

$$\begin{cases} q(t) = \dfrac{t(b_1+b_2)^2 + (b_1+b_2)(b_1-b_2)\sqrt{t(t-1)}}{(b_1-b_2)^2 + 4tb_1b_2}, \\ p(t) = \dfrac{b_1 q(t) - (b_1+b_2)/2}{q(t)\,(q(t)-1)}. \end{cases}$$

We substitute $(q(t), p(t))$ in H_{VI} to get

$$H_{\mathrm{VI}}(b_1, b_2, -1/2, 0; t, q(t), p(t)) = \frac{1}{4t(t-1)}\Big[-b_1 + 2tb_1 + b_1^2 - 2tb_1^2 + 2\sqrt{t(t-1)}b_1^2 \\ - b_2 + 2b_1b_2 + b_2^2 - 2tb_2^2 - 2\sqrt{t(t-1)}b_2^2\Big].$$

Henceforth we denote this formula by $\mathcal{H}_0$.

Starting from $(q(t), p(t))$ and using the translation $\mathbb{C}^4 \to \mathbb{C}^4$, $\mathbf{b} \mapsto \mathbf{b} + n(0,0,1,0)$, we get a series of solutions $\{(q_n(t), p_n(t))\}$. We denote the τ-function associated with $(q_n(t), p_n(t))$ by τ_n.

The Toda equation of the series of solutions of $S_{\mathrm{VI}}(\mathbf{b})$ is

$$\frac{d}{dt} t(t-1) \frac{d}{dt} \log \tau_n + (b_1 + b_3 + n)(b_3 + b_4 + n) = c(n) \frac{\tau_{n-1}\tau_{n+1}}{\tau_n^2}$$

([O1], p. 368, formula (4.4)).

Let

$$\tau_n(t) = \exp\left[\int \left\{\mathcal{H}_0 - \frac{n\,(b_1 t - (b_1+b_2)/2)}{t(t-1)}\right\} dt\right] T_n(t),$$

and substitute it in the Toda equation with $c(n) = 1$ for all n. We get a recurrence formula

(11.1.1) $$T_{n+1} = A_{n+1}/T_{n-1},$$

where

(11.1.2)
$$A_{n+1} = \left\{\frac{1}{4}\left(-2b_1^2 - 2b_2^2 + (b_1^2 - b_2^2)\frac{2t-1}{\sqrt{t(t-1)}}\right) + \left(n - \frac{1}{2}\right)^2\right\} + T_n^2 \\ + (2t-1)T_n' T_n - t(t-1)\left(T_n'' T_n - {T_n'}^2\right).$$

We can also see that its initial condition is $T_0(t) = T_1(t) = 1$.

Since $\sqrt{t(t-1)}$ appears in the denominators of the recurrence formulas (11.1.1) and (11.1.2), $T_n(t)$ is not a polynomial for $n \geq 2$. Introducing, however, a new variable v by

$$v = \sqrt{\frac{t}{t-1}} + \sqrt{\frac{t-1}{t}},$$

we can rewrite the recurrence formula (11.1.1), (11.1.2) as

(11.2.3) $$T_{n+1} = B_{n+1}/T_{n-1},$$

where

(11.2.2)

$$B_{n+1} = \left\{\frac{1}{4}(-2b_1^2 - 2b_2^2) + (b_1^2 - b_2^2)v + \left(n - \frac{1}{2}\right)^2\right\} T_n^2 + \frac{1}{4}(v^2-4)\left\{(v^2-4)\frac{d^2T_n}{dv^2} + v\frac{dT_n}{dv}\right\} T_n - \left\{\frac{1}{2}(v^2-4)\frac{dT_n}{dv}\right\}^2,$$

and $T_n = T_n(b_1, b_2, v)$. We can prove that $T_n(b_1, b_2, v)$ is a polynomial in three variables b_1, b_2 and v (see (17.4)).

12. Generalization of Jacobi polynomials

The sixth Painlevé equation $P_{\rm VI}$ contains the hypergeometric functions. More precisely, the system $S_{\rm VI}(\mathbf{b})$, $\mathbf{b} = (b_1, b_2, b_3, b_4) \in \mathbb{C}^4$, of differential equations, has the solution (q, p) whose τ-function is the hypergeometric function

$$F(a, b, c; t) = 1 + \frac{ab}{1 \cdot c}t + \frac{a(a+1)b(b+1)}{1 \cdot 2 \cdot c(c+1)}t^2 + \cdots \tag{12.1}$$

(Propositon 5.3 in [O1], p. 378), where

$$\begin{aligned} b_1 &= \tfrac{1}{2}(a - b + 1), & b_2 &= -\tfrac{1}{2}(a + b + 1), \\ b_3 &= -\tfrac{1}{2}(a - b + 1), & b_4 &= \tfrac{1}{2}(a + b - 1). \end{aligned} \tag{12.2}$$

Let (q_n, p_n) be the solutions of the system $S_{\rm VI}(\mathbf{b})$ of differential equations obtained by the transformation of the solution (q, p) associated with the translation by $(n-1)(0, 0, 1, 0)$, and τ_n the τ-function associated with (q_n, p_n). Then we can show that $\tau_0 = 1$ and that $\tau_1 = F(a, b, c_1; t)$, up to constant multiplication.

Now, if we assume $c(m) = 1$ for all m, the Toda equation gives the following recurrence formula:

(12.3)

$$\tau_{m+1} = \frac{1}{\tau_{m-1}}\left[(2t-1)\tau_m'\tau_m + t(t-1)\left(\tau_m''\tau_m - {\tau_m'}^2\right) + m(b_3 + b_4 + m)\tau_m^2\right]$$

with initial condition $\tau_0 = 1$, $\tau_1 = F(a, b, c; t)$, as noted above. Particularly, if $\tau_1 = F(a, b, c; t)$ is a polynomial, it is natural to expect (12.3) to generate special polynomials. For example, let n be an integer ≥ 0 and

$$\tau_1 = \text{the } n\text{th Jacobi polynomial},$$

so that

$$\tau_1 = G_n(\alpha, \gamma; t) := F(-n, \alpha + n, \gamma : t),$$

where α, $\gamma \in \mathbb{C}$ are parameters, so that

$$a = -n, \quad b = \alpha + n, \quad c = \gamma. \tag{12.4}$$

The first three Jacobi polynomials are

$$G_0(\alpha,\gamma;t) = 1,$$
$$G_1(\alpha,\gamma;t) = 1 - \frac{\alpha+1}{\gamma}t,$$
$$G_2(\alpha,\gamma;t) = 1 - \frac{\alpha+2}{\gamma}t + \frac{(\alpha+2)(\alpha+3)}{\gamma(\gamma+1)}t^2.$$

Since it is preferable to eliminate denominators, we take

$$T_1(n;e,t;t) := f\,(f+1)\cdots(f+n-1)F(-n,e,f;t)$$

as τ_1. Namely

$$a=-n, \quad b=e, \quad c=f. \tag{12.5}$$

Since we know $b_3+b_4 = b-1 = e-1$ by (12.2) and (12.5), the recurrence formula (12.3) becomes

(12.6)
$$\tau_{m+1} = \frac{1}{\tau_{m-1}}\left[(2t-1)\tau_m'\tau_m + t(t-1)\left(\tau_m''\tau_m - {\tau_m'}^2\right) + m(e+m-1)\tau_m^2\right]$$

with initial condition $\tau_0 = 1$, $\tau_1 = T_1(n;e,f;t)$. We denote τ_m by $T_m(n;e,f;t)$. We can show that $T_m(n;e,f;t)$ is a polynomial in three variables e, f and t. We give some examples. $n=0$ is a trivial but typical case (see (17.6)). For $n=1$,

$$T_1(1;e,f;t) = -tT_1\left(1;f,e,\frac{1}{t}\right),$$
$$T_2(1;e,f;t) = et^2T_1\left(2;f,e,\frac{1}{t}\right),$$
$$T_3(1;e,f;t) = -2e^2(e+1)t^3T_1\left(3;f,e,\frac{1}{t}\right),$$
$$T_4(1;e,f;t) = 12e^3(e+1)^2(e+2)t^4T_1\left(4;f,e,\frac{1}{t}\right).$$

The first nontrivial case appears when $n=2$ (see (17.6)).

13. What are these polynomials?

The Yablonskii-Vorob′ev polynomials and the Okamoto polynomials were discovered in the 1950s and early 1980s, respectively. Being interested in these polynomials and trying to discover more special polynomials, we obtained the results in sections 9–12. The Yablonskii-Vorob′ev polynomials and the Okamoto polynomials have integral coefficients, but since the polynomials in sections 9–12 contain parameters, we thought it would be easier to study the coefficients of their t^l's.

Since it takes about one page to print out the polynomial $T_4(b_1,b_2,v)$ in section 11 and several pages for the polynomial $T_5(b_1,b_2,v)$ (see (17.5.1)), it would be natural to expect a simpler expression. Noumi and Okamoto made the following observation. Let us set

$$z = \frac{2-v}{4}, \quad w = \frac{2+v}{4}, \quad d = -4b_2^2, \quad c = -4b_2^2,$$

$$d_n := (d+1^2)(d+3^2)\cdots\bigl(d+(2n-1)^2\bigr)\,z^n,$$
$$c_n := (c+1^2)(c+3^2)\cdots\bigl(c+(2n-1)^2\bigr)\,w^n.$$

If we put

$$2^{n(n-1)}T_n(b_1,b_2,v) = U_n(c,d,w,z),$$

then U_n satisfies a recurrence formula

(13.1)

$$\begin{aligned}(z+w)U_{n+1}U_{n-1} &= \bigl[cz+dw+(2n-1)^2(z+w)\bigr]\,(w+z)U_n^2\\ &+4zw\left[4zw\left(\frac{\partial}{\partial w}-\frac{\partial}{\partial z}\right)^2 U_n + 2(w-z)\left(\frac{\partial}{\partial w}-\frac{\partial}{\partial z}\right)U_n\right]U_n\\ &-\left[4zw\left(\frac{\partial}{\partial w}-\frac{\partial}{\partial z}\right)U_n\right]^2.\end{aligned}$$

We have

$$U_0 = U_1 = 1,$$
$$U_2 = d_1 + c_1,$$
$$U_3 = d_1d_2 + 3c_1d_2 + 3c_2d_1 + c_1c_2,$$

and

$$\begin{aligned}U_4 = {}& d_1d_2d_3 + 6c_1d_2d_3 + 15c_2d_1d_3 + 10d_1d_2c_3 + 10d_3c_1c_2\\ &+ 15c_1c_3d_2 + 6c_2c_3d_1 + c_1c_2c_3.\end{aligned}$$

So we can express $2T_4 = U_4$ in a line. The suffixes show a distinctive pattern.

For a given integer $n \geq 2$, divide the set $\{1,2,\ldots,n-1\}$ into two disjoint subsets I, J:

$$\{1,2,\ldots,n-1\} = I \cup J, \quad I \cap J = \emptyset,$$

where

$$\begin{cases} I = \{i_1,i_2,\ldots,i_r\}, & i_1 > i_2 > \cdots > i_r,\\ J = \{j_1,j_2,\ldots,j_s\}, & j_1 > j_2 > \cdots > j_s.\end{cases}$$

So we have $r + s = n - 1$.

We denote the product $\prod_{l\in I} c_l \prod_{m\in J} d_m$ by $(i_1,i_2,\ldots,i_r;j_1,j_2,\ldots,j_s)$ or by $(I;J)$. For example, $(\emptyset;3,2,1) = d_1d_2d_3$, $(1;3,2) = c_1d_2d_3$, $(3,1;2) = c_1c_3d_2$. In this notation,

(13.2)
$$\begin{aligned}U_4 = {}&(\emptyset;3,2,1) + 6(1;3,2) + 15(2;3,1) + 10(2,1;3)\\ &+ 10(3;2,1) + 15(3,1;2) + 6(3,2;1) + (3,2,1;\emptyset).\end{aligned}$$

Since

$$(\emptyset;3,2,1) = (d+1^2)(d+3^2)(d+5^2)\cdot(d+1^2)(d+3^2)\cdot(c+1^2)z^6,$$
$$(1;3,2) = (c+1^2)\cdot(d+1^2)(d+3^2)(d+5^2)\cdot(d+1^2)(d+3^2)z^5w,$$

in the right-hand side of (13.2), $(\emptyset;3,2,1)$ represents the monomial involving z^6, $(1;3,2)$ z^5w, $(2;3,1)$ z^4w^2, $(2,1;3)$ and $(3;2,1)$ z^3w^3, $(3,1;2)$ z^2w^4, $(3,2;1)$ zw^5, $(3,2,1;\emptyset)$ w^6, respectively. The coefficients 1, 6, 15, 20 $(= 10+10)$, 15, 6, 1 that appear in (13.2) are binary coefficients, and it is natural that the term of z^3w^3 is divided half-by-half as $10(2,1;3) + 10(3;2,1)$, because the recurrence formula of U_n

is symmetric with respect to transposition of (c, w) and (d, z). Now let us observe the representation of U_5. We have

$$U_5(c, d; z, w) = \sum_{\substack{I \cup J = \{1,2,3,4\} \\ I \cap J = \emptyset}} a_{IJ}(I; J),$$

where a_{IJ} is a positive integer (see (17.5.2)). Calculation by Mathematica reveals the term of $w^7 z^3$ in U_5 as

$$70(3; 4, 2, 1) + 50(2, 1; 4, 3).$$

So what rule assigns the coefficient 70 to the pair of partitions $(3; 4, 2, 1)$ and 50 to $(2, 1; 4, 3)$?

Noumi and Okamoto conjectured that generally we have a representation

$$U_n(c, d; z, q) = \sum_{\substack{I \cup J = \{1,2,3,\ldots,n-1\} \\ I \cap J = \emptyset}} a_{IJ}(I; J),$$

where a_{IJ} is a positive integer. Looking at Noumi's calculation, Soichi Okada conjectured Theorem 13.5 below.

To a given set of integers $I = \{i_1, i_2, \ldots, i_r\}$, $i_1 > i_2 > \cdots > i_r > 0$, we attach the following Young diagram $\lambda(I)$. (1) Put i_1 boxes horizontally on the first row. (2) Put i_1 boxes vertically below the leftmost box of the first row. (3) Put i_2 boxes horizontally and flush-left on the second row (as in Fig. (13.3)). (4) Put i_2 boxes vertically and flush-up on the second column (as in Fig. (13.4)). (5) Iterate procedures 3 and 4 up to i_r to get the Young diagram, $\lambda(I)$.

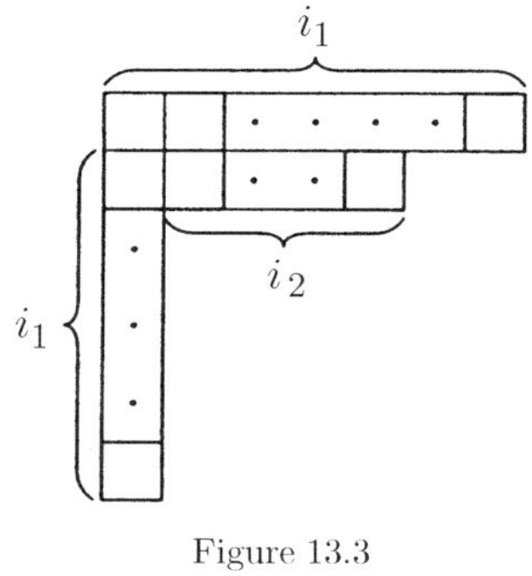

Figure 13.3

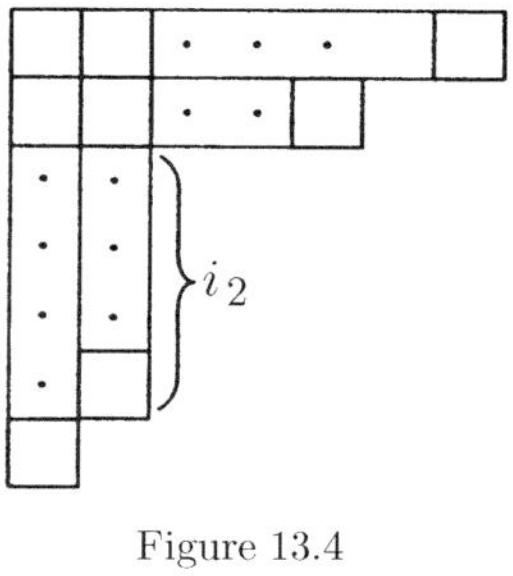

Figure 13.4

THEOREM 13.5. *We have the following representation:*

$$U_n(c, d; z, w) = \sum_{\substack{I \cup J = \{1,2,3,\ldots,n-1\} \\ I \cap J = \emptyset}} \dim L\left(\lambda(I)\right)_{GL_n} (I; J),$$

where $L\left(\lambda(I)\right)_{GL_n}$ is the irreducible representation of GL_n, determined by the Young diagram $\lambda(I)$.

The theorem was proved by Taneda ([T2]). The theorem is an epoch-making result in the study of the Painlevé equations. For, it reveals that there is a relation between the Painlevé equations and combinatorics of the Young diagrams. It is unimaginable from the motivation of the discovery of the Painlevé equations that these two objects are related. It assured us that the Painlevé equations define special functions.

There are similar conjectures, which are not proved, on the other special polynomials generated by the Painlevé equations (see [NOO]).

The proof given by Taneda is, however, based on a direct verification of the recurrence formula (13.1). Everyone would like to prove the theorem in a natural framework. This problem seems to be one of the motivations of Noumi and Yasuhiko Yamada, when they tried to generalize the Painlevé equations from a viewpoint of simple Lie algebras and affine Weyl groups. They are producing a series of interesting results ([NYi], $1 \leq i \leq 4$, [Y]).

14. Yablonskii-Vorob′ev polynomials, Okamoto polynomials and Young diagrams

Actually, the Yablonskii-Vorob′ev and the Okamoto polynomials are also closely related with the Young diagrams, as the $U_n(c, d; z, w)$'s are ([KO2], [NY1]). We explain this by taking the Yablonskii-Vorob′ev polynomials as an example. To this end, we give a brief review of the Schur polynomials.

Let z, x_1, x_2, x_3, ... be variables. The elementary Schur polynomials $S_k(x)$ are defined by the identity

$$\sum_{k \in \mathbb{Z}} S_k(x) z^k = \exp\left(\sum_{k=1}^{\infty} x_k z^k\right).$$

Notice that $S_k(x) = 0$ for $k < 0$, and that $S_1(x) = 1$. More explicitly,

$$S_k(x) = \sum_{k_1 + 2k_2 + \cdots = k} \frac{x_1^{k_1}}{k_1!} \frac{x_2^{k_2}}{k_2!} \cdots$$

for $k > 0$. A descending series of nonnegative integers $\lambda_1 \geq \lambda_2 \geq \cdots \geq \lambda_l \geq 0$ is called a partition and is denoted by $\lambda = (\lambda_1, \lambda_2, \ldots, \lambda_l)$. The number of nonzero λ_i is called the length of the partition λ, and is denoted by $l(\lambda)$.

For a given partition λ, we define the Schur polynomial $S_\lambda(x)$ by

$$S_\lambda(x) = \det\left(S_{\lambda_i - i + j}(x)\right)_{1 \leq i, j \leq l(\lambda)}.$$

For $N \geq l(\lambda)$, representations of GL_N and the Schur polynomials $S_\lambda(x)$ have the following relation. Let π_λ be the irreducible representation of GL_N defined by the partition λ. We have

$$S_\lambda(x) = \mathrm{tr}.\pi_\lambda \begin{bmatrix} \epsilon_1 & & & \\ & \epsilon_2 & & \\ & & \ddots & \\ & & & \epsilon_N \end{bmatrix},$$

where

$$x_j = (\epsilon_1^j + \epsilon_2^j + \cdots + \epsilon_N^j)/j \quad \text{for } j = 1, 2, \ldots.$$

THEOREM 14.1. *Let* $\lambda = (n-1, n-2, \ldots, 1)$ *and* $\tau_n(x) = S_\lambda(x)$. *Then a polynomial* $\bar{T}_n(t)$ *in* t*, obtained by substituting* $x_1 = t$, $x_3 = -4/3$ *and* $x_j = 0$ *for* $j \geq 5$ *in* τ_n *coincides with the Yabonskii-Vorob'ev polynomial defined in section* 6 *up to constant multiplication.*

The proof of this theorem is based on the following principle. According to the theory of soliton equations, the pair of the τ-functions $\tau_n(x)$ and $\tau_{n+1}(x)$ is a solution of a countable number of differential equations called the first modified KP hierarchy ([JM], p. 996). The first two of them are

$$\begin{cases} (D_1^2 + D_2)\tau_n \cdot \tau_{n+1} = 0, \\ (D_1^3 - 4D_3 - 3D_1D_2)\tau_n \cdot \tau_{n+1} = 0, \end{cases} \tag{14.2}$$

where D_j is Hirota's bilinear operator, e.g., the left-hand side of the first equality in (14.2) means

$$\left(\frac{\partial^2}{\partial y_1^2} + \frac{\partial}{\partial y_2}\right) \tau_n(x+y)\tau_{n+1}(x-y)\bigg|_{y=0}. \tag{14.3}$$

On the other hand, it is known that $\tau_n = S_\lambda(x)$ in Theorem 14.1 does not contain x_2, x_4, ... for an arbitrary integer $n \geq 0$, and therefore (14.2) is also written as

$$\begin{cases} D_1^2\tau_n \cdot \tau_{n+1} = 0, \\ (D_1^3 - 4D_3)\tau_n \cdot \tau_{n+1} = 0. \end{cases} \tag{14.4}$$

Here, we recall that if each x_i is defined to be of weight i, the Schur polynomial τ_k is a homogeneous polynomial of weight $k(k-1)/2$. Namely,

$$\left(x_1\frac{\partial}{\partial x_1} + 3x_3\frac{\partial}{\partial x_2} + \cdots\right)\tau_k = \frac{k(k-1)}{2}\tau_k. \tag{14.5}$$

Let $\bar{\tau}_k$ be a polynomial in x_1 and x_3, obtained by substituting $x_j = 0$ $(j \geq 5)$ into τ_k. Then by (14.4) and (14.5), we have

$$\begin{cases} D_1^2\bar{\tau}_n \cdot \bar{\tau}_{n+1} = 0, \\ (D_1^3 - 4D_3)\bar{\tau}_n \cdot \bar{\tau}_{n+1} = 0, \end{cases} \tag{14.6}$$

and

$$x_1\frac{\partial\bar{\tau}_k}{\partial x_1} + 3x_3\frac{\partial\bar{\tau}_k}{\partial x_3} = \frac{k(k-1)}{2}\bar{\tau}_k. \tag{14.7}$$

By (14.6) and (14.7), we have

$$\begin{cases} D_1^2\bar{\tau}_n \cdot \bar{\tau}_{n+1} = 0, \\ \left(D_1^3 - \dfrac{4}{3x_3}(-x_1D_1 - n)\right)\bar{\tau}_n \cdot \bar{\tau}_{n+1} = 0. \end{cases}$$

Therefore, we have

$$\begin{cases} D_t^2\bar{T}_n(t) \cdot \bar{T}_{n+1}(t) = 0, \\ (D_t^3 - tD_t - n)\bar{T}_n(t) \cdot \bar{T}_{n+1}(t) = 0. \end{cases} \tag{14.8}$$

On the other hand, it follows by a simple calculation from (14.8) that $\bar{T}'_{n+1}/\bar{T}_{n+1} - \bar{T}'_n/\bar{T}_n$ satisfies $P_{\rm II}(-n)$ (see remark following proposition 8, [FOU]). As we saw in section 6, for a series of the Yablonskii-Vorob′ev polynomials T_n,

$$T'_{n+1}/T_{n+1} - T'_n/T_n$$

is the rational solution of the second Painlevé equation $P_{\rm II}(-n)$. So we have $T_n = \bar{T}_n$ up to a constant multiplication.

This theorem shows that the Yablonskii-Vorob′ev polynomial can be written as the determinant of a matrix. Many special solutions of the Painlevé equations have a determinant representation ([KO1], [Y]).

15. Discrete Painlevé equations

It is hard to tell which is more essential, continuous or discrete?(2) Since I am an algebraist by origin, I certainly prefer what is discrete. On the other hand, we must recognize that what is discrete is quite difficult to treat. Differential equations play a leading part but difference equations, which are a discrete analogue of differential equations, are also studied.

Henceforth, by a difference equation we mean an algebraic difference equation. A difference equation that has the following two properties is called a discrete Painlevé equation:

(15.1) The difference equation has the property that is a discrete analogue of "being free from movable singularities".

(15.2) A difference equation yields in general a differential equation in a limit. The difference equation reduces to a Painlevé equation in a limit.

As for 15.1, "singularity confinement" is commonly regarded as a discrete analogue of "being free from movable singularities". If a difference equation satisfies the two conditions below, then the equation is said to pass the singularity confinement test.

(15.3) Even if a singularity of a solution dependent on the initial value appears, one can determine the later time-evolution of the solution.

(15.4) A solution preserves the information of its initial value after it passes through a singularity.

Let us see what conditions 15.3 and 15.4 mean, taking an example.

EXAMPLE 15.5. Consider a second-order difference equation

$$x_{n+1} + x_{n-1} = \frac{ax_n}{1 - x_n^2},$$

where a is a nonzero constant. If $x_0 = P$ and $x_1 = 1$, we cannot determine x_2, because the denominator on the right-hand side vanishes. So we set $x_0 = P$ and $x_1 = 1 + \epsilon$ to get

$$\begin{aligned} x_2 &= -\frac{a}{2}\epsilon^{-1} + \frac{-a - 4P}{4} + O(\epsilon), \\ x_3 &= -1 + \epsilon + O(\epsilon^2), \\ x_4 &= -P + O(\epsilon). \end{aligned}$$

This shows that for the initial condition $x_0 = P$ and $x_1 = 1$, x_2 is undetermined but we can determine $x_3 = -1$ and $x_4 = -P$. So condition 15.3 is satisfied. The equality $x_4 = -P$ contains the information on $x_0 = P$ so that condition 15.4 is satisfied too.

We have been thinking for a long time that conditions 15.3 and 15.4 should be paraphrased as follows.

(15.6) An algebraic difference equation represents by definition a birational automorphism of an algebraic variety, and if we take an appropriate model, it induces a biregular automorphism of the model.

Moreover, in most of the cases we can take a projective model (see Example 15.7, Theorems 15.8, 15.10, and Example 15.11). As Example 15.5 shows, the singularity confinement test and algebraic difference equations both belong to birational geometry.

If we look for discrete Painlevé equations under conditions 15.1 and 15.2, we encounter too many equations. It reminds us of the classification of algebraic differential equation $y^{(3)} = R(t, y, y', y'')$ without movable singularities explained in Section 1. For this reason, few specialists in physics and engineering were interested in discrete Painlevé equations. The situation remained unchanged even when Jimbo and Sakai discovered the following discretization of the sixth Painlevé equation in 1995.

EXAMPLE 15.7 ([JS]). For $q \in \mathbb{C}$ and $0 < |q| < 1$, let a_1, a_2, a_3, a_4 and b_1, b_2, b_3, b_4 be parameters satisfying

$$\frac{a_1 a_2}{a_3 a_4} = q \frac{b_1 b_2}{b_3 b_4}.$$

Consider a difference equation

$$q\text{-}P_{\mathrm{VI}} \begin{cases} y_n y_{n+1} = \dfrac{a_3 a_4 (x_{n+1} - tb_1)(x_{n+1} - tb_2)}{(x_{n+1} - b_3)(x_{n+1} - b_4)}, \\ x_n x_{n+1} = \dfrac{b_3 b_4 (y_n - ta_1)(y_n - ta_2)}{(y_n - a_3)(y_n - a_4)}, \end{cases}$$

where t is the independent variable.

Namely, to calculate (x_{n+1}, y_{n+1}) from (x_n, y_n), we first calculate x_{n+1} by the second equation and then y_{n+1} by the first equation. Here we may assume $a_1 a_2 = 1$, $a_3 a_4 = 1$, $b_1 b_2 = q$ and $b_3 b_4 = 1$ by an appropriate change of scale of y, z and t so that q-P_{VI} contains four parameters. We could imagine that $x_{n+1}(t) = x_n(qt)$, $y_{n+1}(t) = y_n(qt)$ so that q-P_{VI} describes the motion of the point $(x_n(t), y_n(t))$ along the discrete time evolution $t \mapsto qt$.

The difference equation q-P_{VI} was obtained by considering a discrete analogue of isomonodromic deformation (see section 2). q-P_{VI} passes the singularity confinement test thanks to the following theorem.

THEOREM 15.8 ([S1]). *Henceforth we shall regard a_i, b_j and q $(1 \le i, j \le 4)$ all as parameters. Consider the fields $K = \mathbb{C}(a_i, b_j, q)_{1 \le i,j \le 4}$ and $L = K(t)$. Then, there exists a projective model M_L of the affine plane $\mathbf{A}^2_L$ such that the K-birational transformation*

$$\Phi^0 \colon A^2_L = L \otimes_K A^2_K \to A^2_L = L \otimes_K A^2_K, \quad (t; x_n, y_n) \mapsto (qt; x_{n+1}, y_{n+1}),$$

induces a K-biregular transformation $\Phi \colon M_L \to M_L$.

Concretely, M_L is obtained by blowing up $\mathbf{P}^1_L \times \mathbf{P}^1_L$ at nine points. This theorem shows that q-P_{VI} describes a discrete motion on the family of projection surfaces M_L parametrized by a_i, b_j, q and t $(1 \le i, j \le 4)$. Moreover, we can prove that the difference equation q-P_{VI} becomes the sixth Painlevé equation P_{VI} as $q \to 1$.

The equation q-P_{VI} is supposed to involve all the formal characters of P_{VI}. In particular, P_{VI} contains four parameters and its Bäcklund transformation group is the extended affine Wely group of type $\tilde{D}_4$. This $\tilde{D}_4$-symmetry is reflected in q-P_{VI}. Moreover, we have the following remarkable result.

THEOREM 15.9 (Gramaticos and Ramani). *q-P_{VI} contains four parameters and the independent variable t. If we regard the independent variable t as a parameter so that q-P_{VI} involves five parameters and if we treat them equally, q-P_{VI} has $\tilde{D}_5$-symmetry.*

This theorem shows that q-P_{VI} has a bigger symmetry than P_{VI}. Namely, the discretization reveals to us a bigger symmetry that we could not see. In other words, when we pass from the discrete Painlevé equations to the classical Painlevé equation by a limit operation, we lose certain information.

Once this theorem is proved, we cannot ignore the discrete Painlevé equations. The algebraic surface M_L in Theorem 15.7 was one of the surfaces studied by Looijenga in 1981 in his research of singularities ([L]), where the $\tilde{D}_5$-symmetry was recognized. Progress of mathematics requires time!

THEOREM 15.10. *q-P_{VI} has a $\tilde{D}_5$-symmetry. Moreover, the symmetry is realized as an automorphism group of the family M_L of projective algebraic surfaces. In other words, we can lift the action of the affine Weyl group G on the rational function field of five variables to a regular action of G on the family M_L of projective surfaces parameterized by a_i, b_j, q, t $(1 \leq i, j \leq 4)$.*

In Theorems 15.9 and 15.10, the motion of the difference equation q-P_{VI}

$$(\mathbf{a}, \mathbf{b}, q, t; x_n, y_n) \mapsto (\mathbf{a}, \mathbf{b}, q, qt; x_{n+1}, y_{n+1})$$

is contained in the $\tilde{D}_5$-symmetry, where $\mathbf{a} = (a_1, a_2, a_3, a_4)$ and $\mathbf{b} = (b_1, b_2, b_3, b_4)$. As this example shows, there is no reason to distinguish the Bäcklund transformations and the discrete motion of a difference equation. More typical is the following example, but we do not go into details.

EXAMPLE 15.11. The difference equation

$$\frac{\alpha + 1/2}{y(\alpha+1) + y(\alpha)} + \frac{\alpha - 1/2}{y(\alpha) + y(\alpha - 1)} = -\left(2y(\alpha)^2 + t\right),$$

obtained by eliminating dy/dt from the Bäcklund transformation

$$y(\alpha+1) = -y(\alpha) - \frac{1 + 2\alpha}{2y(\alpha)^2 + 2\dfrac{dy(\alpha)}{dt} + t}$$

of $P_{\mathrm{II}}(\alpha)$, can be regarded as a discrete first Painlevé equation, where $y(\alpha)$ is a solution of $P_{\mathrm{II}}(\alpha)$ and $y(\alpha+1)$, $y(\alpha-1)$ are the Bäcklund transforms of $y(\alpha)$ by T_+ and T_-, respectively.

This is a motivation for the geometric study of the Bäcklund transformation of the classical Painlevé equations P_{II}, P_{III}, $\ldots$, P_{VI}, and Okamoto's work [O1] attracted new attention. An analogue of Theorem 15.8 also holds for the Bäcklund transformations. The case of P_{II} is as follows:

EXAMPLE 15.12. Using the notation in section 5, $i(b) = -b$, $t_+(b) = b+1$ and $t_-(b) = b - 1$ define the action of the extended affine Weyl group G of type $\tilde{A}_1$ on the field $\mathbb{C}(b)$, and the action is lifted to the K-birational action of G on $\mathbf{A}^2_L$ by

$$(b, t; y, z) \mapsto (-b, t; q + b/p, p),$$

$$(b, t; y, z) \mapsto \left(b+1, t; -q - b/p, -p + 2(q + b/p)^2 + t\right),$$

$$(b, t; y, z) \mapsto \left(b - 1, t; -q + \frac{b-1}{p - 2q^2 - t}, -p + 2q^2 + t\right).$$

There exists a projective model $\mathcal{X}_L$ of $\mathbf{A}^2_L$ such that G is a group of biregular automorphisms of the family $\mathcal{X}_L$ of projective surfaces parametrized by b and t ([U9]).

Research in this direction brought us the following result.

THEOREM 15.13 ([SU]). *The Bäcklund transformations of the Painlevé equations arise from rational double points.*

This theorem answers a long asked question about the origin of the Bäcklund transformations.

Two questions below arise. Noumi and Yamada systematically constructed the following example as a generalization of the Painlevé equations.

EXAMPLE 15.14. Let $\mathfrak{h}$ be a Cartan subalgebra of a simple Lie algebra $\mathfrak{g}$ defined over $\mathbb{C}$. So the extended affine Weyl group of $\mathfrak{g}$ acts on the field $\mathbb{C}(\mathfrak{h})$ of rational functions on $\mathfrak{h}$. Then there exists a field extension M of $\mathbb{C}(\mathfrak{h})$ such that the action of G on $\mathbb{C}(\mathfrak{h})$ can be extended over M ([NY3]).

PROBLEM 15.15. In example 15.14, is it possible to find a projective model $X/\mathbb{C}(\mathfrak{h})$ of the field $M/\mathbb{C}(\mathfrak{h})$ extension such that G acts regularly on X?

The models in Theorems 15.7, 15.9 and Example 15.11 are all two-dimensional. In these cases, the results might depend on the particular characteristics of two-dimensional birational geometry.(3) In Noumi and Yamada's example 15.14, the model generally is of dimension greater than 3.

The group of $\mathbb{C}$-automorphisms of the field $\mathbb{C}(x_1, x_2, \dots, x_n)$ of rational functions in n variables is called the Cremona group of n variables, and is denoted by Cr_n. When we regard a group of Bäcklund transformations as a subgroup of the Cremona group, we can expect that they have some interesting properties.

The Bäcklund transformations

$$(t, b, q, p) \mapsto (t, -b, q + b/p, p),$$

$$(t, b, q, p) \mapsto \big(t, b + 1, -q - b/p, -p + 2(q + b/p)^2 + t\big),$$

denoted respectively by T_- and T_+, of the second Painlevé equation P_{II} generate a subgroup G isomorphic to the affine Weyl group of type $\tilde{A}_1$, of the Cremona group of four variables. For fixed t, we have a subgroup G_t of Cr_3 isomorphic to G.

PROBLEM 15.16. (1) Is the subgroup G_t a maximal subgroup of the Cremona group Cr_3 of three variables? (2) For fixed $t = t_0$, is $G_t \subset Cr_3$ a unique deformation of the subgroup G_{t_0} of Cr_3, up to inner automorphism of Cr_3?

We can pose similar questions on the group of Bäcklund transformations of all the Painlevé equations, all the discrete Painlevé equations, and the generalization of the Painlevé equations by Noumi and Yamada.

16. Differential Galois theory

Differential Galois theory is one of the theories most expected to develop in the 19th century. The theory is essentially infinite-dimensional. In 1887, Picard established the Galois theory of linear ODEs called Picard-Vessiot theory today. His theory is, however, a finite-dimensional theory. Drach's thesis in 1898 is the first attempt at infinite-dimensional differential Galois theory. But this article is a problematic work containing incomplete definitions and wrong proofs. Kolchin's theory is a generalization of Picard-Vessiot theory but his theory is also finite-dimensional. Constructing an infinite-dimensional differential Galois theory was a problem long to be solved. We proposed an infinite-dimensional differential Galois theory as follows. Namely, let L/K be an ordinary differential field extension, the characteristic of the fields being equal to 0. We assume that the field L is finitely generated over K as an abstract field. To the differential field extension L/K, we attach a formal group $Gal(L/K)$ of (generally) infinite dimension ([U5], [U6]). Why is our theory nice? The reason is simple. Because we knew no invariant of the ordinary differential field extension L/K or of the algebraic ordinary differential equation corresponding to the differential field extension L/K. In other words, $Gal(L/K)$ is the unique so far known invariant of the ordinary differential field extension L/K. It is an interesting problem to calculate $Gal(K\langle y\rangle/K)$ of the ordinary differential field extension $K\langle y\rangle/K$ generated over an ordinary differential field K by a solution y of a Painlevé equation. This is, however, a difficult problem to be solved in further developments. If we recall the history of the Galois theory of algebraic equations, we notice the following fact. We could finally solve, by classical Galois theory, historically difficult problems such as the impossibility of solving the general quintic algebraic equation by extraction of radicals and four arithmetic operations. This does not mean, however, that classical Galois theory drew down the curtain and a drama was finished. On the contrary, the discovery of Galois theory is a starting point for further developments in number theory. We hope that differential Galois theory plays a similar role.

17. Tables

(17.1) Painlevé equations

$$P_{\mathrm{I}} \qquad \frac{d^2y}{dt^2} = 6y^2 + t,$$

$$P_{\mathrm{II}} \qquad \frac{d^2y}{dt^2} = 2y^3 + ty + \alpha,$$

$$P_{\mathrm{III}} \qquad \frac{d^2y}{dx^2} = \frac{1}{y}\left(\frac{dy}{dt}\right)^2 - \frac{1}{t}\frac{dy}{dt} + \frac{\alpha y^2 + \beta}{t} + \gamma y^3 + \frac{\delta}{y},$$

$$P_{\mathrm{IV}} \qquad \frac{d^2y}{dt^2} = \frac{1}{2y}\left(\frac{dy}{dt}\right)^2 + \frac{3}{2}y^3 + 4ty^2 + 2(t^2 - \alpha)y + \frac{\beta}{y},$$

$$P_{\mathrm{V}} \qquad \frac{d^2y}{dt^2} = \left(\frac{1}{2y} + \frac{1}{y-1}\right)\left(\frac{dy}{dt}\right)^2 - \frac{1}{t}\frac{dy}{dt} + \frac{(y-1)^2}{t^2}\left(\alpha + \frac{\beta}{y}\right) + \frac{\gamma y}{t} + \frac{\delta y(y+1)}{y-1},$$

$$P_{\mathrm{VI}} \qquad \frac{d^2y}{dt^2} = \frac{1}{2}\left(\frac{1}{y} + \frac{1}{y-1} + \frac{1}{y-t}\right)\left(\frac{dy}{dt}\right)^2 - \left(\frac{1}{t} + \frac{1}{t-1} + \frac{1}{y-t}\right)\frac{dy}{dt} + \frac{y(y-1)(y-t)}{t^2(t-1)^2}\left(\alpha + \frac{\beta t}{y^2} + \frac{\gamma(t-1)}{(y-1)^2} + \frac{\delta t(t-1)}{(y-t)^2}\right),$$

where α, β, γ, $\delta \in \mathbb{C}$ are parameters. See sections 3, 10 and table 1 in [U3] for their Hamiltonian representations.

(17.2) Okamoto polynomials
Recurrence formula:

$$S_n(x) = \frac{1}{S_{n-2}(x)} \left(S''_{n-1}(x) S_{n-1}(x) - S'_{n-1}(x)^2 + (x^2 + 2n - 3) S_n(x)^2 \right).$$

$$S_0(x) = 1,$$

$$S_1(x) = 1,$$

$$S_2(x) = x,$$

$$S_3(x) = 1 + x^2,$$

$$S_4(x) = 5 + 5x^2 + 5x^4 + x6,$$

$$S_5(x) = 175 + 350x^2 + 175x^4 + 140x^6 + 65x^8 + 14x^{10} + x^{12},$$

$$\begin{aligned} S_6(x) &= 67375 + 134750x^2 + 202125x^4 + 107800x^6 + 42350x^8 + 20020x^{10} \\ &\quad + 8050x^{12} + 2200x^{14} + 355x^{16} + 30x^{18} + x^{20}, \\ S_7(x) &= 337211875 + 1011635625x^2 + 1011635625x^4 + 1146520375x^6 \\ &\quad + 838212375x^8 + 371896525x^{10} + 118243125x^{12} + 36295875x^{14} \\ &\quad + 11836825x^{16} + 3570875x^{18} + 854315x^{20} + 147525x^{22} \\ &\quad + 17325x^{24} + 1295x^{26} + 55x^{28} + x^{30}. \end{aligned}$$

(17.3) Special polynomials associated with $P_{\rm III}$
Recurrence formula:

$$\begin{aligned} S_n(a,u) = \frac{1}{S_{n-2}} &\left[-4u^3 \left(\frac{\partial S_{n-1}}{\partial u} + u \frac{\partial^2 S_{n-1}}{\partial u^2} \right) S_{n-1} \right] \\ &\qquad + 4u^4 \left(\frac{\partial S_{n-1}}{\partial u} \right)^2 + (1 + au) S_{n-1}^2. \end{aligned}$$

$$S_0(a,u) = 1,$$

$$S_1(a,u) = 1,$$

$$S_2(a,u) = 1 + au,$$

$$S_3(a,u) = 1 + 3au + 3a^2u^2 + (-2 + a)a(2 + a)u^3,$$

$$\begin{aligned} S_4(a,u) &= 1 + 6au + 15a^2u^2 + 20(-1 + a)a(1 + a)u^3 \\ &\quad + 15(-2 + a)a^2(2 + a)u^4 \\ &\quad + 6(-2 + a)a(2 + a)(-6 + a^2)u^5 \\ &\quad + (-4 + a)(-2 + a)a^2(2 + a)(4 + a)u^6. \end{aligned}$$

(17.4) Special polynomials associated with P_{V}
Recurrence formula:

$$T_n(v,t) = \frac{1}{T_{n-2}}\left[\left(\frac{t}{8} - v + \frac{3(n-1)}{4}\right)T_{n-1}^2 + \frac{\partial T_{n-1}}{\partial t}T_{n-1} + t\frac{\partial^2 T_{n-1}}{\partial t^2}T_{n-1} - t\left(\frac{\partial T_{n-1}}{\partial t}\right)^2\right].$$

$$T_0 = 1,$$

$$T_1 = 1,$$

$$2^3 T_2(v,t) = 6 - 8v + t,$$

$$2^9 T_3(v,t) = 32(1-v)(-5+4v)(-3+4v) + 12(-5+4v)(-3+4v)t + 24(1-v)t^2 + t^3,$$

$$\begin{aligned} 2^{18} T_4(v,t) = {} & 512(-1+v)(-3+2v)(-7+4v)(-5+4v)^2(-3+4v) \\ & + 1536(1-v)(-3+2v)(-7+4v)(-5+4v)(-3+4v)t \\ & + 240(-7+4v)(-5+4v)^2(-3+4v)t^2 \\ & + 80(-5+4v)(-45+80v-32v^2)t^3 \\ & + 480(-1+v)(-3+2v)t^4 + 12(5-4v)t^5 + t^6. \end{aligned}$$

(17.5.1) Special polynomials associated with P_{VI}
Recurrence formula:

$$\begin{aligned} T_n(b_1,b_2,v) = \frac{1}{T_{n-2}}\Bigg[& \left(\frac{1}{4}(-2b_1^2 - 2b_2^2) + (b_1^2 - b_2^2)v + \left(n - \frac{1}{2}\right)^2\right)T_{n-1}^2 \\ & + \frac{1}{4}(v^2-4)\left((v^2-4)\frac{\partial^2 T_{n-1}}{\partial v^2} + v\frac{\partial T_{n-1}}{\partial v}\right)T_{n-1} \\ & - \left(\frac{1}{2}(v^2-4)\frac{\partial T_{n-1}}{\partial v}\right)^2\Bigg]. \end{aligned}$$

$$T_0(b_1,b_2,v) = 1,$$

$$T_1(b_1,b_2,v) = 1,$$

$$4T_2(b_1,b_2,v) = 1 - 2b_1^2 - 2b_2^2 + b_1^2 v - b_2^2 v,$$

$$\begin{aligned} 64T_3(b_1,b_2,v) = {} & 9 - 38b_1^2 + 28b_1^4 - 8b_1^8 - 38b_2^2 + 120b_1^2b_2^2 - 24b_1^4b_2^2 + 28b_2^4 - 8b_2^6 \\ & + 15b_1^2v - 36b_1^4v + 12b_1^6v - 15b_2^2v + 12b_1^4b_2^2v + 36b_2^4v - 12b_1^2b_2^4v \\ & - 12b_2^6v + 15b_1^4v^2 - 6b_1^6v^2 - 30b_1^2b_2^2v^2 + 6b_1^4b_2^2v^2 + 15b_2^4v^2 \\ & + 6b_1^2b_2^4v^2 - 6b_2^6v^2 + b_1^2v^3 - 2b_1^4v^3 + b_1^6v^3 - b_2^2v^3 \\ & - 3b_1^4b_2^2v^3 + 2b_2^4v^3 + 3b_1^2b_2^4v^3 - b_2^6v^3, \end{aligned}$$

$$
\begin{aligned}
4096T_4(b_1, b_2, v) = 2025 &- 13212b_1^2 + 24364b_1^4 - 14656b_1^6 + 4272b_1^8 - 832b_1^{10} + 64b_1^{12} \\
&- 13212b_2^2 + 76468b_1^2b_2^2 - 105280b_1^4b_2^2 + 47552b_1^6b_2^2 - 6720b_1^8b_2^2 \\
&+ 384b_1^{10}b_2^2 + 24364b_2^4 - 105280b_1^2b_2^4 + 66080b_1^4b_2^4 - 16000b_1^6b_2^4 \\
&+ 960b_1^8b_2^4 - 14656b_2^6 + 47552b_1^2b_2^6 \\
&- 16000b_1^4b_2^6 + 1280b_1^6b_2^6 + 4272b_2^8 - 6720b_1^2b_2^8 + 960b_1^4b_2^8 \\
&- 832b_2^{10} + 384b_1^2b_2^{10} + 64b_2^{12} + 5130b_1^2v - 24252b_1^4v \\
&+ 23520b_1^6v - 10656b_1^8v + 2400b_1^{10}v - \cdots .
\end{aligned}
$$

(17.5.2) Polynomials U_n

Recurrence formula:

$$
\begin{aligned}
&(z + w)U_{n+1}U_{n-1} \\
&\quad = \left(cz + dw + (2n-1)^2(z + w)\right)(w + z)U_n^2 \\
&\quad\quad + 4zw\left\{4zw\left(\frac{\partial}{\partial w} - \frac{\partial}{\partial z}\right)^2 U_n + 2(w - z)\left(\frac{\partial}{\partial w} - \frac{\partial}{\partial z}\right)U_n\right\}U_n \\
&\quad\quad - \left\{\left(\frac{\partial}{\partial w} - \frac{\partial}{\partial z}\right)U_n\right\}^2 .
\end{aligned}
$$

$$
\begin{aligned}
U_0(c, d; z, w) &= 1, \\
U_1(c, d; z, w) &= 1, \\
U_2(c, d; z, w) &= (1 + d)w + (1 + c)z, \\
U_3(c, d; z, w) &= (1 + d)^2(9 + d)w^3 + 3(1 + c)(1 + d)(9 + d)w^2z \\
&\quad + 3(1 + c)(9 + c)(1 + d)wz^2 + (1 + c)^2(9 + c)z^3 \\
&\quad + 3(1 + c)(9 + c)(1 + d)wz^2 + (1 + c)^2(9 + c)z^3, \\
U_4(c, d; z, w) &= (1 + d)^3(9 + d)^2(25 + d)w^6 \\
&\quad + 6(1 + c)(1 + d)^2(9 + d)^2(25 + d)w^5z \\
&\quad + 15(1 + c)(9 + c)(1 + d)^2(9 + d)(25 + d)w^4z^2 \\
&\quad + 10(1 + c)(9 + c)(25 + c)(1 + d)^2(9 + d)w^4z^2 \\
&\quad + 10(1 + c)^2(9 + c)(25 + c)(1 + d)(9 + d)w^4z^2 \\
&\quad + 15(1 + c)^2(9 + c)^2(25 + c)(1 + d)(9 + d)w^2z^4 \\
&\quad + 6(1 + c)^2(9 + c)^2(25 + c)(1 + d)wz^5 \\
&\quad + (1 + c)^3(9 + c)^2(25 + c)z^6,
\end{aligned}
$$

$$
\begin{aligned}
U_5(c, d; z, w) &= (1 + d)^4(9 + d)^3(25 + d)^2(49 + d)w^{10} \\
&+ 10(1 + c)(1 + d)^3(9 + d)^3(25 + d)^2(49 + d)w^9z \\
&+ 10(1 + c)(1 + d)^3(9 + d)^3(25 + d)^2(49 + d)w^9z \\
&+ 45(1 + c)(9 + c)(1 + d)^3(9 + d)^2(25 + d)^2(49 + d)w^8z^2
\end{aligned}
$$

$$
\begin{aligned}
&+ 70(1+c)(9+c)(25+c)(1+d)^3(9+d)^2(25+d)(49+d)w^7z^3 \\
&+ 50(1+c)^2(9+c)(1+d)^2(9+d)^2(25+d)^2(49+d)w^7z^3 \\
&+ 35(1+c)(9+c)(25+c)(49+c)(1+d)^3(9+d)^2(25+d)w^6z^4 \\
&+ 175(1+c)^2(9+c)(25+c)(49+c)(1+d)^2(9+d)^2(25+d)(49+d)w^6z^4 \\
&+ 126(1+c)^2(9+c)(25+c)(49+c)(1+d)^2(9+d)^2(25+d)w^5z^5 \\
&+ 126(1+c)^2(9+c)^2(25+c)(1+d)^2(9+d)(25+d)(49+d)w^5z^5 \\
&+ 175(1+c)^2(9+c)^2(25+c)(49+c)(1+d)^2(9+d)(25+d)w^4z^6 \\
&+ 35(1+c)^3(9+c)^2(25+c)(1+d)(9+d)(25+d)(49+d)w^4z^6 \\
&+ 50(1+c)^2(9+c)^2(25+c)^2(49+c)(1+d)^2(9+d)w^3z^7 \\
&+ 70(1+c)^3(9+c)^2(25+c)(49+c)(1+d)(9+d)(25+d)w^3z^7 \\
&+ 45(1+c)^3(9+c)^2(25+c)^2(49+c)(1+d)(9+d)w^2z^8 \\
&+ 10(1+c)^3(9+c)^3(25+c)^2(49+c)(1+d)wz^9 \\
&+ (1+c)^4(9+c)^3(25+c)^2(49+c)z^{10}.
\end{aligned}
$$

(17.6) Generalized Jacobi polynomials
Recurrence formula:

$$
\begin{aligned}
&T_{m+1}(n;e,f;t) \\
&\quad = \frac{1}{T_{m-1}(n;e,f;t)}\big[(2t-1)T'_m(n;e,f;t)T_m(n;e,f;t) \\
&\qquad\qquad + t(t-1)\left\{T''_m(n;e,f;t)T_m(n;e,f;t) - T_m(n;e,f;t)^2\right\} \\
&\qquad\qquad + m(e-1+m)T_m(n;e,f;t)^2\big],
\end{aligned}
$$

where $'$ denotes $\partial/\partial t$. Notice that $T_0(n;e,f;t)$ and $T_1(n;e,f;t)$ are ordinary Jacobi polynomials for fixed $n \geq 0$.

Henceforth we will be seeking $T_m(n;e,f;t)$ for a fixed n and given $T_0(n;e,f;t)$ and $T_1(n;e,f;t)$, by use of the difference equation above.

For $n = 0$,

$$T_0(0;e,f;t) = 1,$$

$$T_1(0;e,f;t) = e,$$

$$T_2(0;e,f;t) = 2e^2(1+e),$$

$$T_3(0;e,f;t) = 12e^3(1+e)^2(2+e),$$

$$T_4(0;e,f;t) = 288e^4(1+e)^3(2+e)^2(3+e).$$

For $n = 1$,

$$
\begin{aligned}
T_0(1; e, f; t) &= 1, \\
T_1(1; e, f; t) &= f - et, \\
T_2(1; e, f; t) &= e\left(f(1+f) - 2(1+e)ft + e(1-e)t^2\right), \\
T_3(1; e, f; t) &= 2e^2(1+e)\big[f(1+f)(2+f) \\
&= 3(2+e)f(1+f)t + 3(1+e)ft^2 - e(1+e)(2+e)t^3\big].
\end{aligned}
$$

For $n = 2$,

$$
\begin{aligned}
T_0(2; e, f; t) &= 1, \\
T_1(2; e, f; t) &= f(1+f) - 2e(1+f)t + e(1+e)t^2, \\
T_2(2; e, f; t) &= e\big[f(1+f)^2(2+f) - e(1+e)(1+f)(2+f)t \\
&\quad + 6(1+e)(1+f)(e+f+ef)t^2 \\
&\quad - 4e(1+e)(2+e)(1+f)t^3 + e(1+e)(2+e)t^4\big], \\
T_3(2; e, f; t) &= 2e^2(1+e)\big[(f(1+f)^2(2+f)^2(3+f) \\
&\quad - 6(2+e)f(1+f)^2(2+f)(3+f)t \\
&\quad + 3(2+e)f(1+f)(2+f)(15+11e+9f+5ef)t^2 \\
&\quad - 4(1+e)(2+e)(1+f)(2+f)(3e+12f+5ef)t^3 \\
&\quad + 3(1+e)(2+e)(1+f)(2+f)(6e+4f+5ef)t^4 \\
&\quad - 6e(1+e)(2+e)^2(3+e)(1+f)t^5 \\
&\quad + e(1+e)^2(2+e)^2(3+e)t^6\big].
\end{aligned}
$$

Notes

(1) For a unified understanding of the Painlevé equations, G should be regarded as the extended affine Weyl group of type $\tilde{A}_1$, although we cannot distinguish the extended affine Weyl group from the affine Weyl group in the case of type $\tilde{A}_1$.

(2) Writing this section, we are grateful to Professor Kenji Kajiwara for teaching us references, history and recent developments of discrete Painlevé equations.

(3) For example, a birational morphism $f\colon S_1 \to S_2$ of a nonsingular projective surface S_1 to another nonsingular projective surface S_2 is reduced to a composite of blow-downs, up to isomorphism.

References

[AL] Arinkin, D. and Lysenko, S., *Isomorphisms between moduli spaces of $SL(2)$-bundles with connections on $P^1 - \{x_1, \ldots, x_4\}$*, Math. Res. Lett. **4** (1997), 181–190.

[C1] Chazy, J., *Sur les équations différentielles du second ordre et du premier degré dont l'intégrale est à points critiques fixes*, Acta Math. **34** (1891), 317–385.

[C2] Chazy, J., *Sur les équations differéntielles dont l'intégrale possède une coupure essentielle mobile*, C. R. Acad. Sci. Paris **150** (1910), 456–458.

[DM] Dubrovin, B. and Mazzocco, M., *Monodromy of certain Painlevé VI transcendents and reflection groups*, Invent. Math. **141** (2000), 55–147.

[Dr] Drach, J., *Essai sur une théorie générale de l'intégration et sur la classification des transcendentes*, Ann. Sci. Ecole Norm. Sup. (3) **15** (1898), 243–384.

[FOU] Fukutani, S., Okamoto, K., and Umemura, H., *Special polynomials and the Hirota bilinear relations of the second and the fourth Painlevé equations*, Nagoya Math. J. **159** (2000), 179–200.

[GNR] Gramaticos, B., Nijhoff, F. W., and Ramani, A., *Discrete Painlevé equations*, The Painlevé Property: One Century Later (R. Conte, ed.), C.R.M. Ser. Math. Phys., Springer-Verlag, New York, 1999, pp. 413–516.

[H] Hadamard, J., *L'œuvre scientifique de Paul Painlevé*, Revue Metaphysique **41** (1934), 289–325.

[H1] Hitchin, N. J., *Twistor spaces, Einstein metrics and isomonodromic deformations*, J. Differential Geometry **42** (1995), 30–112.

[H2] Hitchin, N. J., *Poncelet polygons and the Painlevé equations*, Geometry and Analysis (Bombay, 1992), Tata Inst. Fund. Res. Bombay, 1995, pp. 151–185.

[J] Jacobi, C. G. J., *Über die Differentialgleichung, welcher die Reihen* $1 \pm 2q + 2q^4 \pm 2q^9 - etc.$, $2\sqrt[4]{q} + 2\sqrt[4]{q^9} + 2\sqrt[4]{q^{25}} + etc.$, *Genuge leisten*, Crelles J. **36** (1848), 97–112.

[JM] Jimbo, M. and Miwa, T., *Solitons and infinite dimensional Lie algebra*, Publ. RIMS, Kyoto Univ. **19** (1983), 943–1001.

[JS] Jimbo, M. and Sakai, H., *A q-analog of the sixth Painlevé equation*, Lett. Math. Phys. **38** (1996), 145–154.

[K] Kajiwara, K., *Discrete Painlevé equations; their variety and future*, private note, 1996.

[KO1] Kajiwara, K. and Ohta, K., *Determinant structure of the rational solutions for the Painlevé II equation*, J. Math. Phys. **37** (1996), 4693–4704.

[KO2] Kajiwara, K. and Ohta, K., *Determinant structure of the rational solutions for the Painlevé IV equation*, J. Phys. A **31** (1998), 2431–2446.

[L] Looijenga, E., *Rational surfaces with an anticanonical cycle*, Ann. Math. **114** (1981), 267–322.

[Mn] Manin, Yu. I., *Sixth Painlevé equation, universal elliptic curve and mirror of* P^2, Amer. Math. Soc. Transl. (2) **186** (1998), 131–151.

[MMT] Matano, T., Matumiya, A., and Takano, K., *On some Hamiltonian structures of Painlevé systems II*, J. Math. Soc. Japan **51** (1999), 843–866.

[Mz] Mazzocco, M., *Picard and Chazy solutions to the Painlevé VI equation*, preprint.

[N] Nishioka K., *A note on the transcendency of Painlevé's first transcendent*, Nagoya Math. J. **109** (1988), 63–67.

[No] Noumi, M., *The world of the Painlevé equations*, Sugaku no Tanosimi **9** (1998), 101–116. (Japanese)

[NOO] Noumi, M., Okada, S., Okamoto, K. and Umemura, H., *Special polynomials associated with the Painlevé equations II*, Proc. Taniguchi Symp., 1997, Integrable Systems and Algebraic Geometry, World Scientific, 1998, pp. 349–372.

[NY1] Noumi, M. and Yamada, Y., *Symmetries in the fourth Painlevé equation and the Okamoto polynomials*, Nagoya Math. J. **153** (1999), 53–86.

[NY2] Noumi, M. and Yamada, Y., *Umemura polynomials for the Painlevé V equation*, Phys. Lett. **A247** (1998), 65–69.

[NY3] Noumi, M. and Yamada, Y., *Affine Weyl groups, discrete integrable systems and Painlevé equations*, Comm. Math. Phys. **199** (1998), 281–295.

[NY4] Noumi, M. and Yamada, Y., *Higher order Painlevé equations of type* $A_l^{(1)}$, Funkcial. Ekvac. **41** (1998), 483–503.

[O1] Okamoto, K., *Sur les feuilletages associés aux équations du second ordre à points critiques fixes de P. Painlevé*, Japan. J. Math. **5** (1979), 1–79.

[O2] Okamoto, K., *The Painlevé equations*, Sugaku **32** (1980), 30–43. (Japanese)

[OJ] Okamoto, K., *Studies on the Painlevé equations J* $(\mathrm{I} \le J \le \mathrm{IV})$, I: Ann. Mat. Pura Appl. **146** (1987), 337–381; II: Japan. J. Math. **13** (1987), 47–76; III: Math. Ann. **275** (1986), 221–255; IV: Funkcial. Ekvac. **30** (1987), 305–332.

[P] Picard, E., *Mémoir sur la théorie des fonctions algébriques de deux variables*, J. de Liouville **5** (1889), 135–319.

[S1] Sakai, H., *Studies on q-analog of the Painlevé VI equation*, master thesis at Kyoto Univ., 1996.

[S2] Sakai, H., *Rational surfaces associated with affine root systems and geometry of the Painlevé equations*, Kyoto Univ. preprint, 1999.

[ST] Shioda, T. and Takano, K., *On some Hamiltonian structures of Painlevé systems I*, Funkcial Ekvac. **40** (1997), 271–291.

[SU] Saito, M. and Umemura, H., *Painlevé equations and rational double points*, in preparation.

[T1] Taneda, M., *Remarks on the Yablonskii-Vorob′ev polynomials*, Nagoya Math. J. **159** (2000), 87–111.

[T2] Taneda, M., *A proof of a conjecture associated with an algebraic solution of the sixth Painlevé equation*, submitted to Nagoya Math. J.

[U1] Umemura, H., *On the irreducibility of the first differential equation of Painlevé*, Algebraic Geometry and Commutative Algebra in Honor of Masayoshi Nagata, Kinokuniya and North Holland, 1987, pp. 771–789.

[U2] Umemura, H., *Irreducibility of the Painlevé transcendental functions*, Sugaku Expositions **2** (1989), 231–252.

[U3] Umemura, H., *The Painlevé equations and classical functions*, Sugaku Expositions **11** (1998), 77–100.

[U4] Umemura, H., *Galois theory of algebraic and differential equations*, Nagoya Math. J. **144** (1996), 1–58.

[U5] Umemura, H., *Differential Galois theory of infinite dimension*, Nagoya Math. J. **144**, (1996), 59–135.

[U6] Umemura, H., *Lie-Drach-Vessiot theory, CR-Geometry and Overdetermined Systems*, Advanced Studies in Pure Math., vol. 25, Math. Soc. Japan, Tokyo, 1997, pp. 364–385.

[U7] Umemura, H., *Special polynomials associated with the Painlevé equations* I, Proc. Workshop Painlevé Transcendents, Montreal 1996, to appear.

[U8] Umemura, H., *Irreducibility of the Painlevé equations, Evolution in the past 100 years*, ibid.

[U9] Umemura, H., *On the transformation group of the second Painlevé equations*, Nagoya Math. J. **157** (2000), 15–46.

[UW1] Umemura, H. and Watanabe, F., *Solutions of the second and fourth Painlevé equations*, I, Nagoya Math. J. **148** (1997), 151–198.

[UW2] Umemura, H. and Watanabe, F., *Solutions of the third Painlevé equation* I, Nagoya Math. J. **151** (1998), 1–24.

[Y] Yamada, Y., *Determinant formulas for the τ-functions of the Painlevé equation of type A*, Nagoya Math. J. **156** (1999), 123–134.

Graduate School of Polymathematics, Nagoya University, Nagoya 4648602, Japan

Translated by HUMIHIKO WATANABE, TOSHINOBU SHIMIZU, and HIROSHI UMEMURA

Amer. Math. Soc. Transl.
(2) Vol. **204**, 2001

The Navier-Stokes Equation in Various Function Spaces

Masao Yamazaki

Dedicated to the memory of late Professor Tosio Kato

ABSTRACT. As an example of the close relationship between real analysis and nonlinear partial differential equations, the author surveys the recent work on the Navier-Stokes equations employing various function spaces, in particular, the work by Miyakawa on the decay of weak solutions, and that of Kozono and Yamazaki on the Cauchy problem with distributions as initial data.

Fourier analysis is, by nature, closely related to partial differential equations. In fact, Fourier originated Fourier analysis in order to analyze the heat equations. One of the most fundamental objects in modern Fourier analysis, namely, the notion of singular integral operators, was introduced by Calderón and Zygmund [**12**] in order to show the uniqueness of the solution to the Cauchy problem for a class of partial differential equations. However, many works on Fourier analysis in the early seventies, especially the epoch-making paper by Fefferman and Stein [**25**] on Hardy spaces, introduced a number of new ideas and methods with which Fourier analysis should be reorganized. Furthermore, these works yielded a number of new problems which became the central issue. As a result, the percentage of works on Fourier analysis directly related to partial differential equations decreased.

The situation has drastically changed in the nineties. Namely, there appeared many deep results on Fourier analysis with powerful application to partial differential equations. Evidence of this is the Fields medal awarded to Bourgain in 1994. There are other important works, by Kenig, Ponce, Vega and others on the smoothing effect of dispersive partial differential equations, by Coifman, Lions, Meyer and Semmes [**19**] on the relationship between the Hardy spaces and compensated compactness, and by Evans and Müller [**23**] on the estimate of solutions to the Poisson

This article originally appeared in Japanese in Sūgaku **51** (3) (1999), 291–308.

2000 *Mathematics Subject Classification.* Primary 35Q30; Secondary 35K55, 42B25, 42B30, 46E30, 46E35.

Key words and phrases. Navier-Stokes equation, Cauchy problem, Hardy spaces, distributions, Besov spaces, Morrey spaces, semilinear heat equations.

Partly supported by Grant-in-Aid for Scientific Research (C) 11640156, Ministry of Education, Sciences, Sports and Culture, Japan.

equation in the local Hardy space and its application to the two-dimensional Euler equation describing the motion of non-viscous incompressible fluids.

As we said above, there are remarkable applications to nonlinear partial differential equations. For one thing, Fourier analysis has now been sufficiently reorganized to provide new powerful tools applicable to nonlinear partial differential equations. For another thing, although the theory of nonlinear partial differential equations at first made rapid progress with fairly simple tools, it has now matured so that sharp estimates based on deep results in Fourier analysis are required for its further progress.

I was invited to write a survey on the above-mentioned recent relationship between Fourier analysis and partial differential equations, but it is a *mission impossible* for me to cover such a wide and important theme. Hence, as an example of such relationships, I would like to concentrate on the study of the Navier-Stokes equation in various function spaces, together with the work by Miyakawa on the decay of the weak solutions of the Navier-Stokes initial value problem, and the joint work by Kozono and myself on the solvability of the initial value problems with as large initial data as possible. I hope other topics will be discussed in forthcoming articles by more appropriate authors.

For general results on harmonic analysis needed here, see the survey article by Arai [**1**], and for general results on the Navier-Stokes equations, see Shibata [**83**] and Wiegner [**95**]. Arai [**1**] refers to partial differential equations closely related to harmonic analysis, which are different from the ones considered in this article.

1. Decay of weak solutions to the Navier-Stokes equation

In this section we survey the results of Miyakawa [**63, 64, 65**] on the decay properties of weak solutions of the Navier-Stokes initial value problem. We first introduce the Hardy spaces and the homogeneous Besov spaces, and then state the main results.

1.1. Function spaces. Basic properties of the Hardy spaces on $\mathbb{R}^n$, together with their history, are surveyed in Arai [**1**]. For further details, see Stein [**86**] for example. Since we employ also the weak Hardy spaces, we give the definition of these spaces.

For p such that $0 < p \leq \infty$, let L^p denote the set of p-th power Lebesgue integrable functions on $\mathbb{R}^n$. Although it is customary to consider this space only for $1 \leq p \leq \infty$, we consider p less than 1 too, since we treat functions decaying faster than the elements of L^1. Then the space L^p becomes an (F)-space (complete and metrizable topological vector space) with quasi-norm

$$\|u\,|L^p\| = \begin{cases} \left(\displaystyle\int_{\mathbb{R}^n} |u(x)|^p dx\right)^{1/p} & \text{for } 0 < p < \infty, \\ \operatorname{ess.sup}|u(x)| & \text{for } p = \infty. \end{cases}$$

Moreover, for $\tilde{p} = \min\{1, p\}$, the quasi-norm $\|\cdot\,|L^p\|$ satisfies the inequality

$$\|u + v\,|L^p\|^{\tilde{p}} \leq \|u\,|L^p\|^{\tilde{p}} + \|v\,|L^p\|^{\tilde{p}}.$$

In particular, for $1 \leq p \leq \infty$, the space L^p becomes a Banach space with norm $\|\cdot\,|L^p\|$.

Next, for $0 < p < \infty$, we introduce the set $L^{p,\infty}$, which we call the set of weak L^p functions on $\mathbb{R}^n$, as the set of Lebesgue measurable functions $u(x)$ on $\mathbb{R}^n$ such that

$$\|u\,|L^{p,\infty}\| = \sup_{t>0} t\mu\left(\left\{x \in \mathbb{R}^n \mid |u(x)| > t\right\}\right)^{1/p} < \infty.$$

Here $\mu(A)$ denotes the Lebesgue measure on $\mathbb{R}^n$. Then the space $L^{p,\infty}$ becomes an (F)-space with quasi-norm $\|\cdot\,|L^{p,\infty}\|$, and contains L^p as a proper subspace. For $a > 0$, the function $|x|^{-a}$ on $\mathbb{R}^n$ does not belong to L^p for any p, but belongs to the space $L^{n/a,\infty}$. Since the fundamental solution of the Laplacian on $\mathbb{R}^n$ is written in the above form for $n \geq 3$, the space L^p naturally appears in the theory of partial differential equations. Furthermore, if $1 < p < \infty$, then there exists a norm on $L^{p,\infty}$ equivalent to $\|\cdot\,|L^{p,\infty}\|$, and the space $L^{p,\infty}$ becomes a Banach space with this norm. On the other hand, both the space $L^{p,\infty}$ for $0 < p \leq 1$ and the space L^p for $0 < p < 1$ have trivial dual spaces. This implies that these spaces can never become Banach spaces, and cannot be contained in the space of distributions. Hence, from the viewpoint of functional analysis, these spaces are not suitable for the study of partial differential equations. The space L^1 has no such functional analytic defects. However, in the study of the Navier-Stokes equation, we often employ a direct sum decomposition called the Helmholtz decomposition in order to remove the pressure term. For the whole space $\mathbb{R}^n$, this decomposition can be represented by virtue of the Riesz transforms, and is not a bounded operator on L^1. Hence it is very difficult to consider the Navier-Stokes equation on $\mathbb{R}^n$ in the framework of L^1.

As substitutes for these spaces, we employ the Hardy spaces and weak Hardy spaces, which have much better properties from the viewpoint of functional analysis and Fourier analysis. In the following let $\mathcal{S}$ and $\mathcal{S}'$ denote the set of rapidly decreasing functions and the set of tempered distributions on $\mathbb{R}^n$ respectively, and fix $\psi \in \mathcal{S}$ with $\int_{\mathbb{R}^n} \psi(x)dx = 1$. Then, for $u \in \mathcal{S}'$, we define the maximal function $M_\psi u$ by the formula

$$(M_\psi u)(x) = \sup_{t>0} \frac{1}{t^n} \left| \left(\psi\left(\frac{\cdot}{t}\right) * u\right)(x)\right| = \sup_{t>0} \left| \int_{\mathbb{R}^n} \psi(y)u(x-ty)dy \right|.$$

Here $u * v$ denotes the convolution of u and v on $\mathbb{R}^n$.

For $0 < p < \infty$, we introduce the Hardy space H^p and the weak Hardy space $H^{p,\infty}$ on $\mathbb{R}^n$ by

$$H^p = \{u \in \mathcal{S}' \mid \|u\,|H^p\| = \|M_\psi u\,|L^p\| < \infty\}$$

and

$$H^{p,\infty} = \{u \in \mathcal{S}' \mid \|u\,|H^{p,\infty}\| = \|M_\psi u\,|L^{p,\infty}\| < \infty\}$$

respectively, by virtue of the maximal function $\mathcal{M}_\psi u$ above. Then the spaces H^p and $H^{p,\infty}$ become (F)-spaces and, in particular, H^p [resp. $H^{p,\infty}$] becomes a Banach space for $1 \leq p < \infty$. [resp. for $1 < p < \infty$.] These spaces are independent of the choice of the function $\psi \in \mathcal{S}$ up to the equivalence of norms. The inclusion relation $H^p \subset H^{p,\infty}$ holds trivially.

The inclusion relation $H^p \subset L^p$ holds for $1 \leq p < \infty$, and the relation $H^{p,\infty} \subset L^{p,\infty}$ for $1 < p < \infty$. On the other hand, if $u \in L^\infty$, then we immediately have $M_\psi u \in L^\infty$, and if $u \in L^1$, then we have $M_\psi u \in L^{1,\infty}$. Hence, in view of the Marcinkiewicz interpolation theorem (see Arai [**1**]), the assumption $u \in L^p$ implies

$M_\psi u \in L^p$ for $1 < p < \infty$. In the same way, in view of Hunt's interpolation theorem, the assumption $u \in L^{p,\infty}$ implies $M_\psi u \in L^{p,\infty}$ for $1 < p < \infty$. It follows that $H^1 \subset L^1 \subset H^{1,\infty}$, and the spaces H^p and $H^{p,\infty}$ coincide with L^p and $L^{p,\infty}$ respectively for $1 < p < \infty$.

On the other hand, the space H^p for $0 < p < 1$ and the space $H^{p,\infty}$ for $0 < p \leq 1$ contain distributions other than measurable functions. For example, the distribution $\delta(x-1) - \delta(x)$ on $\mathbb{R}$ belongs to the space H^p for $1/2 < p < 1$, and to the space $H^{p,\infty}$ for $1/2 \leq p \leq 1$, where $\delta(x)$ denotes the Dirac δ-function on $\mathbb{R}$. We thus see that there are no inclusion relations between the spaces L^p and H^p for $0 < p < 1$, and no relations between $L^{p,\infty}$ and $H^{p,\infty}$ for $0 < p \leq 1$.

Next, in order to define the Besov spaces, we introduce the homogeneous Littlewood-Paley decomposition on $\mathbb{R}^n$. Let $\chi(t)$ be a monotone-decreasing smooth function on $(0, +\infty)$ satisfying the inequality $0 \leq \chi(t) \leq 1$ and the identities $\chi(t) \equiv 1$ on $(0, 4/3]$ and $\chi(t) \equiv 0$ on $[3/2, +\infty)$, and define a function $\varphi_j(\xi)$ on $\mathbb{R}^n$ by $\varphi_j(\xi) = \chi\left(2^{-j}|\xi|\right) - \chi\left(2^{1-j}|\xi|\right)$ for every $j \in \mathbb{Z}$. Then $\varphi_j(\xi)$ becomes a smooth function on $\mathbb{R}^n$ satisfying the relation $\operatorname{supp} \varphi_j \subset \left\{\xi \in \mathbb{R}^n \mid 2^{j+1}/3 \leq |\xi| \leq 2^{j-1}3\right\}$ and the identity $\varphi_j(\xi) \equiv \varphi_0\left(2^{-j}\xi\right)$. Furthermore, the summation $\sum_{j=-\infty}^{\infty} \varphi_j(\xi)$ is locally finite and a partition of unity on the set $\mathbb{R}^n \setminus \{0\}$. If $u \in \mathcal{S}'$ satisfies $\operatorname{supp} \mathcal{F}[u](\xi) = \{0\}$, then the structure theorem of distributions implies that $\mathcal{F}[u]$ can be represented as a linear combination of finite order derivatives of the Dirac δ-function on $\mathbb{R}^n$, and it follows that u is a polynomial on $\mathbb{R}^n$. Here $\mathcal{F}$ denotes the Fourier transform on $\mathbb{R}^n$, and $\mathcal{F}^{-1}$ denotes the inverse transform. Hence, for every class $u \in \mathcal{S}'/\mathcal{P}$, there exists a unique sequence in $\mathcal{S}'$ indexed by $j \in \mathbb{Z}$ defined as $\{\varphi_j(D)u\}_{j=-\infty}^{\infty}$, where $\mathcal{P}$ denotes the set of polynomials on $\mathbb{R}^n$. Here, in view of the fact that we have $\mathcal{F}^{-1}[P(\xi)\mathcal{F}[u](\xi)](x) = P(D)u$ for every $P(\xi) \in \mathcal{P}$ with

$$D = (D_1, \ldots, D_n) = \left(-i\frac{\partial}{\partial \xi_1}, \ldots, -i\frac{\partial}{\partial \xi_n}\right),$$

we put $\varphi(D)u = \mathcal{F}^{-1}[\varphi(\xi)\mathcal{F}[u](\xi)]$ for general functions $\varphi(\xi)$.

Let p and r be positive numbers, and let s be a real number. For $u(x) \in \mathcal{S}'/\mathcal{P}$, put

$$\left\| u \,\middle|\, \dot{B}^s_{p,r} \right\| = \left\| \left\{ 2^{sj} \left\| \varphi_j(D)u(x) \,|\, L^p \right\| \right\}_{j=-\infty}^{\infty} \,\middle|\, \ell^r \right\|.$$

Using this, define the space $\dot{B}^s_{p,r}$ as

$$\dot{B}^s_{p,r} = \left\{ u \in \mathcal{S}'/\mathcal{P} \mid \left\| u \,\middle|\, \dot{B}^s_{p,r} \right\| < \infty \right\}.$$

This space becomes an (F)-space, and if p, $r \in [1, \infty]$, then this space becomes a Banach space. This space is independent of the choice of the function φ_0 in the Littlewood-Paley decomposition up to the equivalence of norms. Roughly speaking, the index s describes the differentiability, and the index p describes the local integrability and the decay order at infinity. The index r makes little difference (if $r < r'$ then it follows that $\dot{B}^s_{p,r} \subset \dot{B}^s_{p,r'}$), and hence is the least important one. Nevertheless, it may play an important role in sharp arguments. Note that here the index s does not mean "differentiability up to the order of s", but "differentiability exactly of the order of s". Hence, there is no inclusion relation between the spaces $\dot{B}^s_{p,r}$ and $\dot{B}^t_{p,r}$ when $s \neq t$.

See Triebel [**89, 90**] for a detailed history and properties of these spaces. In order to treat the case $0 < p < 1$ too, these books give complicated computations using the maximal function $M_\psi u$ introduced above. However, for applications to partial differential equations, it usually suffices to limit ourselves to the cases $1 \le p \le \infty$. For these cases, we can easily deduce the properties of these spaces only from simple properties of Radon measures, and hence it is more elementary than Fourier analysis on the usual L^p-spaces, which is essentially based on the boundedness of singular integrals on these spaces. (See Bergh and Löfström [**6**, Chapter 6].) It is to be noted that these books mainly treat inhomogeneous Besov spaces $B^s_{p,r}$. This space corresponds to the inhomogeneous Littlewood-Paley decomposition, namely, the partition of unity on $\mathbb{R}^n$ in which the origin is covered by the support of only one function. In this context, the index s means differentiability up to order s, and hence $s > t$ implies $B^s_{p,\infty} \subset B^t_{p,r}$. The theory of these spaces is simpler in the sense that we need not consider equivalence classes modulo $\mathcal{P}$. However, the theory of time-global solvability of nonlinear partial differential equations is essentially based on the estimates of the decay of solutions, to describe which we must consider homogeneous function spaces. (If we limit ourselves to time-local solutions, we can discuss only inhomogeneous spaces, and the treatment is easier.)

In spite of the necessity described above, it is awkward to identify modulo polynomials in the application to partial differential equations. In many cases we can get rid of this awkwardness by the following remark.

REMARK 1. Suppose that either $s < n/p$ or $s = n/p$, $r \le 1$. Then, for every $u \in \dot{B}^s_{p,r}$, the finite sum $\sum_{j=-M}^{N} \varphi_j(D)u$ converges in the topology of $\mathcal{S}'$ as $M, N \to \infty$, and the limit belongs to the original equivalence class u modulo polynomials. By taking the limit above as the canonical representative of each equivalence class, we can regard $\dot{B}^s_{p,r}$ as a subspace of $\mathcal{S}'$ in these cases. (See Bourdaud [**10**].)

It follows from this remark and the Frobenius theorem on the complete integrability of Pfaffian systems in the space of distributions that, for every positive integer m, we can regard $\dot{B}^s_{p,r}$ as a subspace of $\mathcal{S}'/\mathcal{P}^{m-1}$ if either $s < m + n/p$ or $s = m + n/p$, $r \le 1$. Here $\mathcal{P}^{m-1}$ denotes the space of polynomials on $\mathbb{R}^n$ of order less than or equal to $m - 1$.

As the dual to the result above, every element $u \in L^1 \cap \dot{B}^s_{p,r}$ satisfies the identity $\int_{\mathbb{R}^n} u(x)dx = 0$ if either $s < n/p - n$ or $s = n/p - n$, $r < \infty$. In the same way, every element $u \in \dot{B}^s_{p,r}$ such that $\int_{\mathbb{R}^n} (|x| + 1)^m |u(x)|dx < \infty$ satisfies the identity $\int_{\mathbb{R}^n} P(x)u(x)dx = 0$ for every polynomial $P(x)$ of order less than or equal to m, provided either $s < n/p - n - m$ or $s = n/p - n - m$, $r < \infty$. We call this property of the space $\dot{B}^s_{p,r}$ the moment vanishing condition.

The norms of the homogeneous function spaces introduced above, as opposed to those of inhomogeneous spaces, are homogeneous, as follows: If $u \in \dot{B}^s_{r,s}$, then for every $\lambda > 0$, the distribution u_λ defined by $u_\lambda(x) = U(\lambda x)$ also belongs to the space $\dot{B}^s_{p,r}$, and the estimate

$$C^{-1}\lambda^{s-n/p} \left\| u \,\middle|\, \dot{B}^s_{p,r} \right\| \le \left\| u_\lambda \,\middle|\, \dot{B}^s_{p,r} \right\| \le C\lambda^{s-n/p} \left\| u \,\middle|\, \dot{B}^s_{p,r} \right\|$$

holds with a constant C independent of $u \in \dot{B}^s_{p,r}$ and $\lambda > 0$. Between spaces with the same homogeneity as above, there exists an inclusion relation, called an embedding theorem of Sobolev type. Namely, if $p < q$, then we have the inclusion relation $\dot{B}^s_{p,r} \subset \dot{B}^{s-n/p+n/q}_{q,r}$.

The Hardy space H^p and the weak Hardy space $H^{p,\infty}$ can be described as function spaces similar to the spaces above. Namely, we have the relations

$$\|u\,|H^p\| \sim \left\| \left\| \{\varphi_j(D)u(x)\}_{j=-\infty}^{\infty} \right| \ell^2 \right\| \left| L^p \right\|$$

and

$$\|u\,|H^{p,\infty}\| \sim \left\| \left\| \{\varphi_j(D)u(x)\}_{j=-\infty}^{\infty} \right| \ell^2 \right\| \left| L^{p,\infty} \right\|.$$

Furthermore, real interpolation yields the embeddings $H^p \subset \dot{B}^{n/q-n/p}_{q,p}$ and $H^{p,\infty} \subset \dot{B}^{n/q-n/p}_{q,\infty}$ for p and q such that $p < q$. (See Jawerth [**39**].) In particular, functions in the Hardy spaces and the weak Hardy spaces satisfy the moment vanishing condition. Namely, if a measurable function $u(x)$ such that $\int_{\mathbb{R}^n}(1+|x|)^m|u(x)|dx < \infty$ belongs to $H^{p,\infty}$ for some $p < n/(n+m)$, or belongs to $H^{n/(n+m)}$, then the identity $\int_{\mathbb{R}^n} P(x)u(x)dx = 0$ holds for every polynomial $P(x)$ of order less than or equal to m.

Conversely, let $u(x)$ be a measurable function on $\mathbb{R}^n$ so that $(1+|x|)^{m+n+1}u(x)$ is bounded on $\mathbb{R}^n$ and the identity $\int_{\mathbb{R}^n} P(x)u(x)dx = 0$ holds for every polynomial $P(x)$ of order less than or equal to m. Then $u(x)$ belongs to $H^{n/(n+m+1),\infty}$ and H^p for every $p > n/(n+m+1)$.

1.2. Decay of weak solutions. We consider the Navier-Stokes equation, which is used as a model of motion of incompressible viscous fluids, on $\mathbb{R}^n$ for $n \geq 2$ as follows:

$$\frac{\partial u}{\partial t} - \Delta_x u + (u \cdot \nabla_x)u - \nabla_x \pi = 0 \quad \text{in } (0,\infty)\times\mathbb{R}^n, \tag{1.1}$$

$$\nabla_x \cdot u = 0 \quad \text{in } (0,\infty)\times\mathbb{R}^n, \tag{1.2}$$

$$u(0,x) = a(x) \quad \text{on } \mathbb{R}^n. \tag{1.3}$$

Here $u(t,x) = (u_1(t,x), \ldots, u_n(t,x))$ is an n-dimensional unknown real vector function standing for the velocity of the fluid at time t and position x, and $\pi(t,x)$ is an unknown real scalar function standing for the pressure of the fluid at t and x. Since π can be obtained modulo constants at once if u is obtained, we usually remove π and consider the equation with u as the only unknown function. We assume that the Cauchy data a is an n-dimensional real-valued vector function satisfying the compatibility condition $\nabla \cdot a = 0$.

It is well known since the pioneering work of Leray [**55**] that the Cauchy problem above admits a time-global weak solution for $a(x) \in L^2$. Furthermore, at least one weak solution $u(t,x)$ satisfies the decay property $\|u(t,\cdot)\,|L^2\| \to 0$ as $t \to +\infty$, as has been verified by Masuda [**60**], Kato [**42**], Schonbek [**77, 78**], Kajikiya and Miyakawa [**40**], and Wiegner [**94**]. However, for $n \geq 3$, neither the uniqueness nor the smoothness of weak solutions is known.

In general, if the initial data $a(x) \in L^2$ satisfy some condition on the spatial decay, then the decay order of weak solutions is explicitly described, and is the same as the decay order of the solution of the linear Stokes equation with the same initial data. Moreover, if the assumption on the spatial decay of the initial data gets stronger, then the decay order of the solutions gets faster accordingly. However, an assertion of this type is valid only until the decay order of the linear equation arrives at some limit where it coincides with the decay order of terms yielded by nonlinear terms, and decay estimates better than the limit above do not hold in general. Results of this type have been obtained by Schonbek [**79, 80**], Schonbek, Schonbek and Süli [**81**] on the whole space, and by Borchers and Miyakawa [**7, 8**] on exterior domains. Recently, Miyakawa [**63, 64, 65**] studied the limit of the decay order of a solution on the whole space by using the function spaces introduced above, and his result seems to be almost optimal. We shall survey his results in the following. For p such that $n/(n+1) \leq p \leq 1$, we define the space X^p as follows:

$$X^p = \begin{cases} \dot{B}^{-1}_{1,\infty} & \text{for } p = \dfrac{n}{n+1}, \\ \dot{B}^{-n/p+n}_{1,1} & \text{for } \dfrac{n}{n+1} < p < 1, \\ H^1 & \text{for } p = 1. \end{cases} \tag{1.4}$$

In view of the embedding properties stated at the end of the preceding subsection, these spaces seem to be the smallest naturally defined Banach spaces containing

$$\begin{cases} H^{n/(n+1),\infty} & \text{for } p = \dfrac{n}{n+1}, \\ H^p & \text{for } \dfrac{n}{n+1} < p \leq 1. \end{cases}$$

This fact can also be understood from the following result on the duality:

$$(H^p)^* = (X^p)^* = \begin{cases} \dot{B}^{n/p-n}_{\infty,\infty} & \text{for } \dfrac{n}{n+1} < p < 1, \\ BMO & \text{for } p = 1, \end{cases} \tag{1.5}$$

together with the real interpolation properties

$$(H^{p_0}, H^{p_1})_{\theta,\infty} = H^{p,\infty} \quad \text{for } p_0 \neq p_1,\ 1/p = (1-\theta)/p_0 + \theta/p_1,$$

$$\left(\dot{B}^{s_0}_{1,1}, \dot{B}^{s_1}_{1,1}\right)_{\theta,\infty} = \dot{B}^{s}_{1,\infty} \quad \text{for } s_0 \neq s_1,\ s = (1-\theta)s_0 + \theta s_1.$$

Here, for $f(x) \in L^1_{\rm loc}(\mathbb{R}^n)$ and for a cube B with all edges parallel to one of the coordinate axes, we put

$$f_B = \frac{1}{|B|}\int_B f(x)dx,$$

$$\|f\,|BMO\| = \sup_B \frac{1}{|B|}\int_B |f(x) - f_B|\,dx,$$

and define the space BMO by

$$BMO = \left\{f(x) \in L^1_{\rm loc}(\mathbb{R}^n)/\mathbb{C} \mid \|f\,|BMO\| < \infty\right\}.$$

Then the main result of [**65**] is as follows:

THEOREM 1. *Suppose that $n/(n+1) \leq p \leq 1$. Then, for every $a(x) \in L^2 \cap H^1 \cap X^p$, there exists a weak solution $u(t,x)$ of (1.1)–(1.3) satisfying the estimate $\|u(t,\cdot)\,|X^r\| \leq C(1+t)^{-n/2p+n/2r}$ for every r such that $n/(n+1) < r \leq 1$ and $p \leq r$. In particular, turbulent solutions in the sense of Leray [**55**], namely weak solutions that satisfy*

$$\left\|u(t,\cdot)\,|L^2\right\|^2 + 2\int_s^t \left\|\nabla u(\tau,\cdot)\,|L^2\right\|^2 d\tau \leq \left\|u(s,\cdot)\,|L^2\right\|^2 \quad \textit{for all } t \geq s$$

for $s = 0$ and for almost all $s > 0$, satisfy the estimate above.

We remark that the existence of turbulent solutions was proved by Leray [**55**] for $\mathbb{R}^3$ and by Kato [**42**] for $\mathbb{R}^4$.

Roughly speaking, this theorem asserts that weak solutions of the Navier-Stokes equation decay faster than the fundamental solution of the heat equation as $t \to +\infty$, provided the spatial decay order of the Cauchy data is sufficiently large. This follows from the condition $\nabla_x \cdot u(t,\cdot) \equiv 0$ and Theorem 2 on the nonlinear term.

Making use of the fact that the weak solution in question is a solution in the sense of distributions of the integral equation obtained by rewriting the Navier-Stokes equation (1.1)–(1.3), we can prove Theorem 1 by estimating the norm of weak solutions. In the estimate we use the duality (1.5) and the estimates

$$\left\| \frac{\partial E}{\partial x_j}(t,\cdot) \,\middle|\, H^p \right\| \le C t^{-(n+1)/2+n/2p} \quad \text{for } \frac{n}{n+1} < p \le 1, \tag{1.6}$$
$$\left\| \frac{\partial E}{\partial x_j}(t,\cdot) \,\middle|\, H^{n/(n+1),\infty} \right\| \le C$$

on the derivatives of the fundamental solution $E(t,x)$ of the heat equation with respect to each coordinate variable x_j. The inequality (1.6) was first used by Carpio [**18**] for $p=1$. As well as these facts, essential use is made of the following theorem, which was shown to be closely related to compensated compactness and applied to various problems in nonlinear partial differential equations by Coifman, Lions, Meyer and Semmes [**19**].

THEOREM 2. *Suppose that $p, q \in (1,\infty)$ and that $1/r = 1/p + 1/q < (n+1)/n$. If two functions $u(x) = \big(u_1(x),\dots,u_n(x)\big) \in (L^p)^n$ and $v(x) = \big(v_1(x),\dots,v_n(x)\big) \in (L^q)^n$ satisfy the condition*

$$\operatorname{div} u(x) = \sum_{j=1}^{n} \frac{\partial u_j}{\partial x_j}(x) \equiv 0,$$
$$\operatorname{rot} v(x) = \left((-1)^{j+k} \left\{ \frac{\partial v_k}{\partial x_j}(x) - \frac{\partial v_j}{\partial x_k} \right\} \right)_{k<j} \equiv 0,$$

then we have $u \cdot v(x) = \sum_{j=1}^{n} u_j(x) v_j(x) \in H^r$, and there exists a constant C such that $\|u \cdot v \,|H^r\| \le C \,\|u\,|L^p\|\,\|v\,|L^q\|$. If $1/p + 1/q = 1 + 1/n$, then the conclusion holds if H^r is replaced by $H^{n/(n+1),\infty}$.

In fact, since $\operatorname{div}\{u_k(x)\}_{k=1}^n = 0$ and since $\operatorname{rot}\big\{\big(\partial u_j/\partial x_k\big)(x)\big\}_{k=1}^n = 0$ for every $j = 1,2,\dots,n$, we can apply Theorem 2 to

$$(u \cdot \nabla_x) u_j(x) = \sum_{k=1}^{n} u_k(x) \cdot \big(\partial u_j/\partial x_k\big)(x)$$

for every $j = 1,2,\dots,n$.

Suppose that the following stationary Navier-Stokes equation on $\mathbb{R}^n$ for $n \ge 3$ with time-independent external force,

$$-\Delta w + \big(w\cdot\nabla\big) w - \nabla\varphi = f \quad \text{in } \mathbb{R}^n, \tag{1.7}$$
$$\nabla \cdot w = 0 \quad \text{in } \mathbb{R}^n, \tag{1.8}$$

admits a solution $w(x)$ with $w(x) = O(|x|^{-1})$ as $|x| \to \infty$. Miyakawa [**64, 65**] considered a small initial perturbation $a(x)$ on the stationary solution $w(x)$ above and obtained similar estimates of the solution $u(t,x)$ of the equation

$$\frac{\partial u}{\partial t} - \Delta_x u + (w \cdot \nabla_x)u + (u \cdot \nabla_x)w + (u \cdot \nabla_x)u - \nabla_x \pi = 0 \qquad \text{in } (0,\infty) \times \mathbb{R}^n, \tag{1.9}$$

$$\nabla_x \cdot u = 0 \qquad \text{in } (0,\infty) \times \mathbb{R}^n, \tag{1.10}$$

$$u(0,x) = a(x) \qquad \text{on } \mathbb{R}^n, \tag{1.11}$$

which describes the evolution of the perturbation to $w(x)$. See [**64, 65**] for details.

Results closely related to the above were also obtained by Carpio [**18**]. Namely, assuming either $a(x) \in L^1 \cap L^2$ or $a(x) \in L^2$ and $(1+|x|)a(x) \in L^1$, Carpio [**18**] obtained the decay order of suitable weak solutions and showed that the solutions can be represented as the sum of a linear equation with $a(x)$ as the Cauchy data and a term that decays faster than the solution above. The assumptions in [**18**] and in [**63, 64, 65**] do not imply each other.

On the other hand, the situation is completely different on exterior domains, namely, the complements of bounded closed subsets of $\mathbb{R}^n$ with smooth boundary. In fact, for three-dimensional exterior domains, Kozono [**47**] showed that the Navier-Stokes equation admits a solution in the class $C\big((0,T), L^1\big)$ only if the net force exerted on the boundary is equal to 0 at every time. This condition is also related to the difference between the decay order for the whole space obtained by [**79, 80**] and that for the exterior domain obtained by [**7, 8**].

As another application of Theorem 2 for the Navier-Stokes equation we have the following. Beale, Kato and Majda [**5**] and Ponce [**76**] showed that, if a solution of the Navier-Stokes equation or the Euler equation on $\mathbb{R}^n$ blows up at a finite time, then the vorticity and the deformation tensor of the solution become unbounded in the space L^∞ near the blow-up time. Kozono and Taniuchi [**48, 49**] recently showed, by using Theorem 2, that the results above are still valid if L^∞ is replaced by the somewhat larger space BMO.

2. The Navier-Stokes equation with distributions as initial data

In this section we consider strong solutions of (1.1)–(1.3). Kato [**42**] proved the existence of time-local solutions for every Cauchy data $a(x) \in L^n$, and that the solution exists globally in time provided the norm of $a(x)$ is sufficiently small. Here we aim to consider (1.1)–(1.3) for $a(x) \notin L^n$; for example, Radon measures and other distributions. In particular, we hereby obtain forward self-similar solutions. First we recall previous results on forward self-similar solutions and function spaces which appear in the works on the Cauchy problem (1.1)–(1.3) with $a(x) \notin L^n$. Then we introduce function spaces that contain the spaces above and state the main results.

This problem is of interest not only in its own right. For one thing, there are fluid dynamical models that do not fall in the framework of L^p-spaces, as will be explained later.

For another, we may be able to find solutions with "bad" properties. In order to show that the weak solutions that he constructed are not classical in general, Leray [**55**] asked whether there exist nontrivial backward self-similar solutions of the Navier-Stokes equation. Nečas, Růžička and Šverák [**67**] answered this question negatively by showing that such solutions never belong to L^n. More recently, Tsai [**91**] showed that there are no non-constant backward self-similar solutions

$u(t,x)$ with $(T,0)$ as the singular point satisfying the local condition

$$\sup_{T-\delta<t<T} \int_{|x|<r} |u(t,x)|^2 dx + \int_{T-\delta}^{T} \int_{|x|<r} |\nabla_x u(t,x)|^2 dx\, dt < \infty \tag{2.1}$$

for sufficiently small δ, $r>0$. On the other hand, Majda [**58**] and Okamoto [**70, 71**] showed the existence of backward self-similar solutions that do not satisfy (2.1) and grow linearly as $|x| \to \infty$. Hence, in very large function spaces, there may well exist backward self-similar solutions decaying slowly at infinity.

As another possibility, there may exist nontrivial stationary solutions that decay at infinity. Even if such stationary solutions are not in standard function spaces, non-stationary solutions in standard spaces may be close to such solutions during time evolution. Then the non-stationary solutions move very slowly. This phenomenon may be observed as turbulence. Hence the analysis of such stationary solutions is also of interest.

Furthermore, there is a chance to find the most suitable function spaces in the course of considering the Cauchy problem in many function spaces. For example, the results of Novotny and Padula [**69**], Borchers and Miyakawa [**9**], and Kozono and Yamazaki [**53, 54**] on the unique existence and the stability of stationary solutions of the Navier-Stokes exterior problem, together with the result of Hishida [**38**] on the Boussinesq equation in exterior domains, imply that the Lorentz space $L^{n,\infty}$ is more suitable to these problems than the standard space L^n. This fact is explained in Shibata [**83**] in detail.

2.1. Forward self-similar solutions of the Navier-Stokes equation. First, in the Cauchy problem (1.1)–(1.3), apply the rescaling of variables

$$\begin{aligned} u(t,x) &\to u_\lambda(t,x) = \lambda u(\lambda^2 t, \lambda x), \\ \pi(t,x) &\to \pi_\lambda(t,x) = \lambda^2 \pi(\lambda^2 t, \lambda x), \\ a(x) &\to a_\lambda(x) = \lambda a(\lambda x) \end{aligned} \tag{2.2}$$

for a positive constant λ. Then the pair (u_λ, π_λ) is a solution of the Cauchy problem (1.1)–(1.3) with a_λ as the Cauchy data. Solutions invariant under this rescaling are said to be forward self-similar solutions. (This is different from backward self-similar solutions in Leray [**55**], discussed above. Forward self-similar solutions can be constructed easily, as we shall see later. Recently, Márek, Nečas, Pokorný, and Schonbek [**59**] introduced the notion of pseudo self-similar solutions, which generalize self-similar solutions, and discussed both types of self-similar solutions in a unified manner.) If we can show that the Cauchy problem (1.1)–(1.3) admits a unique time-global solution for a class of the Cauchy data containing functions $a(x)$ that satisfy the scale invariance $\lambda a(\lambda x) = a(x)$, then we can construct forward self-similar solutions, and the Cauchy data of forward self-similar solutions necessarily enjoy this scale invariance. However, nontrivial functions $a(x)$ with this scale invariance do not belong to any L^p-spaces. Hence, to obtain forward self-similar solutions, it is necessary to consider the Cauchy problem (1.1)–(1.3) in function spaces strictly larger than L^p.

The study of self-similar solutions is partly motivated by the asymptotic behavior of the solutions in the case $n=2$. Namely, in order to study the asymptotic profile for large (t,x) of the solutions $\big(u(t,x), \pi(t,x)\big)$ of (1.1)–(1.3) in the case $n=2$, we consider the behavior of the rescaled pair $(u_\lambda(t,x), \pi_\lambda(t,x))$ in (2.2) as $\lambda \to \infty$. Applying the operator rot to the Cauchy data $a_\lambda(x)$, we obtain the initial vorticity $\omega_\lambda(x) = \lambda^2\omega(\lambda x)$, which converges to $\Omega\delta$ in the sense of weak convergence of Radon measures as $\lambda \to +\infty$. Here we set $\Omega = \int_{\mathbb{R}^n} \omega(x)dx$. On the other hand, the function $a(x)$ satisfying $\operatorname{rot} a(x) = \Omega\delta$ is of the form $c(-x_2/|x|^2, x_1/|x|^2)$, and the solution with initial data of this form is a forward self-similar solution.

Historically, Cottet [**20**] and Giga, Miyakawa and Osada [**34**] first studied (1.1)–(1.3) for $n = 2$ with Radon measures as initial vorticity, in order to describe the time evolution of two-dimensional flow with a finite number of vortices localized at single points. (Kato [**44**] improved these results later on.) These results were applied to the study of the asymptotic behavior of such solutions by Giga and Kambe [**31**] and Carpio [**17**]. Okamoto [**71**] constructed explicitly a number of forward self-similar solutions for $n = 3$. Tian and Xin [**88**] constructed some families of rotationally self-similar solutions for $n = 3$. A survey of these results and an elementary method to construct solutions are given in Giga and Giga [**30**, Chapter 2].

Furthermore, Miyakawa and Yamada [**66**] and Michaux and Rakotoson [**61**] considered similar problems on bounded domains with boundary conditions different from the Dirichlet condition.

Then, inspired by the study of Niwa [**68**] on the Cauchy problem with Radon measures as initial data for semilinear heat equations of Fujita type in the form

$$\frac{\partial u}{\partial t} = \Delta_x u + |u|^{\gamma-1}u \tag{2.3}$$

(see Fujita [**26**]), Giga and Miyakawa [**33**] considered (1.1)–(1.3) for $n = 3$ with initial vorticity in the spaces of Radon measures of Morrey type, in order to treat the problem in the fluid dynamics of time-evolution of flows with vortex strings as initial vorticity. Then they gave sufficient conditions for the unique existence of time-local solutions and of time-global solutions. Similar problems are also treated by Cottet and Soler [**21**]. Then Taylor [**87**], Kato [**43**] and Federbush [**24**] studied (1.1)–(1.3) with initial data itself in the Morrey spaces or the spaces of Radon measures of Morrey type for general space dimension n.

On the other hand, Kobayashi and Muramatu [**46**], Grubb [**36**], Cannone [**14, 15**], Cannone and Planchon [**16**], Planchon [**74, 75**], and Vishik [**92**] studied (1.1)–(1.3) in the Besov spaces. Barraza [**3, 4**] and Kato and Ponce [**45**] also studied (1.1)–(1.3) with initial data in weak L^p spaces and $\dot{H}^s_p$ with $s < 0$ respectively.

2.2. Function spaces. In this subsection we construct function spaces that contain the function spaces of the preceding subsection and on which the Cauchy problem (1.1)–(1.3) admits a unique time-global solution. The function spaces contain distributions other than Radon measures as well. Namely, our function spaces contain both the Morrey spaces and the Besov spaces. Our construction is similar to that of the Besov spaces, but as the basis we employ the Morrey spaces instead of the standard L^p-spaces.

We first introduce the Morrey spaces. Let p and q satisfy the assumption $1 \le q \le p < \infty$, and put

$$\|u\,|\mathcal{M}_{p,q}\| = \sup_{a\in\mathbb{R}^n, R>0} R^{n/p-n/q}\left(\int_{B(a,R)} |u(x)|^q dx\right)^{1/q} \tag{2.4}$$

for $u(x) \in L^q_{\mathrm{loc}}$, where $B(a,R)$ denotes the closed ball in $\mathbb{R}^n$ with center a and radius R. By using this norm we introduce the space

$$\mathcal{M}_{p,q} = \{u(x) \in L^q_{\mathrm{loc}} \mid \|u\,|\mathcal{M}_{p,q}\| < +\infty\}.$$

Further, for p such that $1 \leq p < \infty$, put

$$\|\nu \,|\mathcal{M}_p\| = \sup_{a\in\mathbb{R}^n, R>0} R^{n/p-n}|\nu|\big(B(a,R)\big)$$

for a Radon measure $\nu(\cdot)$, where $|\nu|(\cdot)$ denotes the total variation of $\nu(\cdot)$. By using this norm we introduce the space

$$\mathcal{M}_p = \{u(x) \mid \|u\,|\mathcal{M}_p\| < +\infty\}.$$

Then $\mathcal{M}_{p,q}$ and $\mathcal{M}_p$ become Banach spaces, and the Radon-Nikodým theorem implies that $\mathcal{M}_{p,1}$ is a closed subspace of $\mathcal{M}_p$.

Our notation is adapted from that of Giga and Miyakawa [**33**] and Taylor [**87**]. The space $\mathcal{M}_{p,q}$ coincides with $L^{q,n-nq/p}$ in Campanato [**13**], which is the standard notation in Fourier analysis, and coincides with $\mathcal{E}^{q,-n/p}$ in Peetre [**73**]. The last notation seems to be theoretically superior, since the index $-n/p$ represents the homogeneity of the norm of this space, as is seen by the equality $\|u_\lambda\,|\mathcal{M}_{p,q}\| = \lambda^{-n/p}\|u\,|\mathcal{M}_{p,q}\|$. But we use this notation, since it appeals to the intuition of researchers in partial differential equations, as we shall see later.

Roughly speaking, the index p of the standard L^p-spaces represents both the local regularity and the decay order at infinity. The smoothness is better if p is larger, and the decay is better if p is smaller. The set $\mathcal{M}_{p,q}$ consists of functions with L^q-like smoothness and L^p-like decay properties. In fact, for $r \leq q < p$, we have the inclusion relation

$$\mathcal{M}_{p,p} = L^p \subset L^{p,\infty} \subset \mathcal{M}_{p,q} \subset \mathcal{M}_{p,r}.$$

The following proposition gives a property peculiar to the Morrey spaces in the sense that the usual L^p-spaces never enjoy this property. This property is used in Giga and Miyakawa [**33**] to analyze the motion of vortex strings.

PROPOSITION 3. *Suppose that $1 \leq m < n$, and write $x = (x_1,\dots,x_n)$ as $x = (x', x'')$, where $x' = (x_1,\dots,x_m)$ and $x'' = (x_{m+1},\dots,x_n)$. Let $u(x)$ denote a function $u(x') \in \mathcal{M}_{p,q}(\mathbb{R}^m)$ regarded as a function of $x \in \mathbb{R}^n$. Then $u(x)$ belongs to $\mathcal{M}_{np/m,q}(\mathbb{R}^n)$, and*

$$\big\|u(x)\,\big|\mathcal{M}_{np/m,q}(\mathbb{R}^n)\big\| \leq C\,\|u(x')\,|\mathcal{M}_{p,q}(\mathbb{R}^m)\|.$$

Next we introduce function spaces of Besov type based on the Morrey spaces, following Kozono and Yamazaki [**50**]. Suppose that real numbers p, q and s satisfy the condition $1 \leq q \leq p < \infty$, suppose that $1 \leq r \leq \infty$, and let $\{\varphi_j(x)\}_{j=-\infty}^{+\infty}$ be the homogeneous Littlewood-Paley partition of unity introduced in subsection 1.1. Then we put

$$\big\|u\,\big|\mathcal{N}^s_{p,q,r}\big\| = \Big\|\big\{2^{sj}\,\|\varphi_j(D)u\,|\mathcal{M}_{p,q}\|\big\}_{j=-\infty}^{\infty}\Big|\,\ell^r\Big\|$$

for $u \in \mathcal{S}'/\mathcal{P}$, and by using this norm we introduce the function space

$$\mathcal{N}^s_{p,q,r} = \big\{u \in \mathcal{S}'/\mathcal{P} \mid \big\|u\,\big|\mathcal{N}^s_{p,q,r}\big\| < \infty\big\}.$$

Then the spaces $\mathcal{N}^s_{p,q,r}$ enjoy the following properties.

PROPOSITION 4. *The following inclusions hold between the spaces $\mathcal{N}^s_{p,q,r}$.*

1. *If $r \leq r'$, then $\mathcal{N}^s_{p,q,r} \subset \mathcal{N}^s_{p,q,r'}$.*
2. $\mathcal{N}^s_{p,q,1} \subset (-\Delta)^{-s/2}\mathcal{M}_{p,q} \subset \mathcal{N}^s_{p,q,\infty}$.
3. $\mathcal{N}^s_{p,1,1} \subset (-\Delta)^{-s/2}\mathcal{M}_{p,1} \subset (-\Delta)^{-s/2}\mathcal{M}_p \subset \mathcal{N}^s_{p,1,\infty}$.

4. *If* $q' < q < p < \infty$, *then* $\mathcal{N}^s_{p,p,r} = \dot{B}^s_{p,r} \subset \mathcal{N}^s_{p,q,r} \subset \mathcal{N}^s_{p,q',r}$.

In particular, these spaces are larger than the homogeneous Besov spaces in general, and they coincide if $q = p$.

We also have the following embedding theorem of Sobolev type, similar to that for the Besov spaces.

PROPOSITION 5. *If* $1 \leq q \leq p < \infty$ *and* $0 < \theta < 1$, *then*

$$\mathcal{N}^s_{p,q,r} \subset \mathcal{N}^{s-(1-\theta)n/p}_{p/\theta,q/\theta,r} \subset \dot{B}^{s-n/p}_{\infty,r}.$$

This proposition and Remark 1 imply that, if either $s < n/p$ or $s = n/p$, $r = 1$, then we can take the canonical representative of each element of $\mathcal{N}^s_{p,q,r}$, namely an equivalence class modulo $\mathcal{P}$. Hence we can regard this space as a subspace of $\mathcal{S}'$ by identifying each equivalence class to its canonical representative.

We furthermore have the following proposition, corresponding to Proposition 3.

PROPOSITION 6. *In the same situation as in Proposition* 3, *let* $u(x)$ *denote a function* $u(x') \in \mathcal{N}^s_{p,q,r}(\mathbb{R}^m)$ *regarded as a function of* $x \in \mathbb{R}^n$. *Then* $u(x) \in \mathcal{N}^s_{np/m,q,r}(\mathbb{R}^n)$, *and*

$$\left\| u(x) \left| \mathcal{N}^s_{np/m,q,r}(\mathbb{R}^n) \right\| \leq C \left\| u(x') \left| \mathcal{N}^s_{p,q,r}(\mathbb{R}^m) \right\| .\right.\right.$$

Differential operators and Fourier multipliers have boundedness properties on these spaces. This fact is the generalization of these boundedness properties on homogeneous Besov spaces.

PROPOSITION 7. *Suppose that* $1 \leq q \leq p < \infty$, $1 \leq r \leq \infty$ *and* $s, \sigma \in \mathbb{R}$. *If a function* $P(\xi) \in C^{[n/2]+1}(\mathbb{R}^n \setminus \{0\})$ *satisfies the estimate*

$$\left| \frac{\partial^{|\alpha|} P}{\partial \xi^\alpha}(\xi) \right| \leq C |\xi|^{\sigma - |\alpha|}$$

for every $\alpha \in \mathbb{N}^n$ *such that* $|\alpha| \leq [n/2] + 1$, *then the operator* $P(D)$ *is bounded from* $\mathcal{N}^s_{p,q,r}$ *to* $\mathcal{N}^{s-\sigma}_{p,q,r}$.

As an application of this proposition, we can prove the following estimate for the heat kernel.

PROPOSITION 8. *Suppose that real numbers* p, q *and* s *are such that* $1 \leq q \leq p$. *Then the Laplacian* Δ *on* $\mathbb{R}^n$ *generates a "bounded analytic semigroup"* $\exp(t\Delta)$ *on* $\mathcal{N}^s_{p,q,r}$ *for every* $r \in [1, \infty]$. *Furthermore, for every real number* σ *greater than* s, *there exists a positive constant* C *such that*

$$\left\| \exp(t\Delta) u \left| \mathcal{N}^\sigma_{p,q,1} \right\| \leq C t^{(s-\sigma)/2} \left\| u \left| \mathcal{N}^s_{p,q,\infty} \right\|\right.\right.$$

for every $u \in \mathcal{N}^s_{p,q,\infty}$.

REMARK 2. The domain of the operator $-\Delta$ as a closed operator in the Banach space $\mathcal{N}^s_{p,q,r}$ is dense if and only if $p = q$ and $1 \leq r < \infty$. It follows that the semigroup $\exp(t\Delta)$ is not a C_0-semigroup on $\mathcal{N}^s_{p,q,r}$ if $q < p$ or $r = \infty$. (This is why we said "bounded analytic semigroup" in Proposition 8.) This implies that, in the Cauchy problem (1.1)–(1.3), the initial condition (1.3) cannot be attained in the usual sense in general, and should be regarded as some sort of weak form.

EXAMPLE 1. Since the Dirac δ-function $\delta(x)$ on $\mathbb{R}$ is a finite Radon measure, it follows that $\delta(x) \in \mathcal{M}_1 \subset \mathcal{N}^0_{1,1,\infty}$ in view of Proposition 4. Hence Proposition 7 implies that $\mathrm{p.v.}\,\frac{1}{x} \in \mathcal{N}^0_{1,1,\infty}$, since it is written as the Hilbert transform of $\delta(x)$ multiplied by a constant. (Observe that $\mathrm{p.v.}\,\frac{1}{x} \notin \mathcal{M}_1$.) Hence, applying Proposition 6, we see that $\mathrm{p.v.}\,\frac{1}{x_1}$, regarded as a distribution on $\mathbb{R}^n$, belongs to $\mathcal{N}^0_{n,1,\infty}$. Now Proposition 5 implies that $\mathrm{p.v.}\,\frac{1}{x_1} \in \mathcal{N}^{n/p-1}_{p,p/n,\infty}$ for every $p \geq n$.

The norm of the space $\mathcal{N}^s_{p,q,r}$ enjoys the following homogeneity property:

$$C^{-1}\lambda^{s-n/p} \left\| u \,|\mathcal{N}^s_{p,q,r} \right\| \leq \left\| u_\lambda \,|\mathcal{N}^s_{p,q,r} \right\| \leq C\lambda^{s-n/p} \left\| u \,|\mathcal{N}^s_{p,q,r} \right\| .$$

Hence the so-called Fujita-Kato principle in [**27**] (see Ozawa [**72**]) implies that the equality $s - n/p = -1$ is necessary for the Cauchy problem (1.1)–(1.3) to admit a time-global solution for all $a(x) \in \mathcal{N}^s_{p,q,r}$ with small norm. In the next subsection we show that this time-global solvability really holds.

2.3. The Navier-Stokes equation with distributions as initial data. We have the following theorem for the Cauchy problem (1.1)–(1.3).

THEOREM 9 (uniqueness). *Suppose that $n \geq 2$, and suppose that real numbers p and q are such that $n \leq p$ and $1 \leq q \leq p$. Then there exists a positive constant M_1 such that the following holds. For every $T \in (0, \infty]$ and every $a \in (\mathcal{S}')^n$ satisfying $\nabla_x \cdot a = 0$, there exists at most one solution $u(t,x)$ of the system (1.1)–(1.2) on $(0,T) \times \mathbb{R}^n$ satisfying the following conditions:*

1. *For every $T' \in (0,T)$, we have*

$$\sup_{0<t\leq T'} \left\| u(t,\cdot) \left| \mathcal{N}^{n/p-1}_{p,q,\infty} \right. \right\| < \infty$$

and

$$\sup_{0<t\leq T'} t^{1/2-n/4p} \left\| u(t,\cdot) \,|\mathcal{M}_{2p,2q} \right\| < \infty.$$

2. *We have*

$$\limsup_{t\to+0} t^{1/2-n/4p} \left\| u(t,\cdot) \,|\mathcal{M}_{2p,2q} \right\| < M_1.$$

3. *As $t \to +0$, we have $u(t,\cdot) \to a$ in the sense of tempered distributions. The initial condition (1.3) holds in this meaning.*

THEOREM 10 (existence of global solutions). *Let n, p, q and M_1 be the same as in Theorem 9. Then there exist a positive constant δ_1 and a strictly monotone-increasing continuous function $\omega_1(\delta)$ defined on $[0,\delta_1]$ satisfying $\omega_1(0) = 0$ and $\omega_1(\delta_1) \leq M_1$ such that the following holds. For every $a \in (\mathcal{N}^{n/p-1}_{p,q,\infty})^n$ satisfying $\|a|\mathcal{N}^{n/p-1}_{p,q,\infty}\| < \delta_1$ and $\nabla_x \cdot a = 0$, there exists a solution $u(t,x)$ of the system (1.1)–(1.2) on $(0,\infty) \times \mathbb{R}^n$ satisfying the following conditions:*

1. *$u(t,x) \in C^\infty((0,\infty) \times \mathbb{R}^n)$.*
2. *$\sup_{t>0} \|u(t,\cdot)|\mathcal{N}^{n/p-1}_{p,q,\infty}\| < \infty$.*
3. *$\sup_{t>0} t^{1/2-n/4p} \left\| u(t,\cdot) \,|\mathcal{M}_{2p,2q} \right\| \leq \omega_1 \left(\|a|\mathcal{N}^{n/p-1}_{p,q,\infty}\| \right)$.*
4. *As $t \to +0$, we have $u(t,\cdot) \to a$ in the weak-$*$ topology of the space $\dot{B}^{-1}_{\infty,\infty}$. The initial condition (1.3) holds in the sense.*

REMARK 3. Both $u(t, \cdot)$ and a belong to $\mathcal{N}^{n/p-1}_{p,q,\infty}$, and they are contained in $\dot{B}^{-1}_{\infty,\infty}$ in view of Proposition 5. The space $\dot{B}^{-1}_{\infty,\infty}$ is regarded as the space dual to $\dot{B}^{1}_{1,1}$.

EXAMPLE 2. In view of Example 1, we see that the distribution

$$a(x) = c\left(0, \cdots, 0, \text{p. v.}\, \frac{1}{x_1}\right)$$

with $c \in \mathbb{R}$ satisfies the assumptions of Theorem 10 provided that $|c|$ is sufficiently small. This is a distribution that is not a Radon measure.

Kato [**43**] showed that the Cauchy problem (1.1)–(1.3) admits a unique time-local solution for sufficiently small $a(x) \in \mathcal{M}_n$. Theorems 9 and 10 generalize the above result, since Propositions 4 and 5 imply that $\mathcal{M}_n \subset \mathcal{N}^0_{n,1,\infty} \subset \mathcal{N}^{n/p-1}_{p,p/n,\infty}$.

Next, suppose that the initial vorticity $\nabla_x \times a(x)$ is a sufficiently small element of $\mathcal{M}_{n/2}$. Since Propositions 4 and 5 imply $\mathcal{M}_{n/2} \subset \mathcal{N}^0_{n/2,1,\infty} \subset \mathcal{N}^{n/p-2}_{p,2p/n,\infty}$ for every $p > n/2$, it follows from the facts $\nabla_x \cdot a(x) = 0$ and $\nabla_x \times a(x) \in \mathcal{M}_{n/2}$ that $a \in \mathcal{N}^{n/p-1}_{p,2p/n,\infty}$, in view of the Biot-Savart law and Proposition 7. Hence we see that Theorems 9 and 10 also hold in this case. This result was proved by Cottet [**20**] and Giga, Miyakawa and Osada [**34**] for $n = 2$, and by Giga and Miyakawa [**33**] for $n = 3$. However, the existence of time-global solutions with large data for $n = 2$ does not follow from our theorems.

Furthermore, if $p > n$, Propositions 4 and 5 imply the inclusion relations $L^{n,\infty} \subset \mathcal{M}_{n,nq/p} \subset \mathcal{N}^0_{n,nq/p,\infty} \subset \mathcal{N}^{n/p-1}_{p,q,\infty}$ and $\dot{B}^{n/p-1}_{p,\infty} = \mathcal{N}^{n/p-1}_{p,p,\infty} \subset \mathcal{N}^{n/p-1}_{p,q,\infty}$ for every $q < p$. Hence we see that our function spaces contain all the function spaces that have been proved, by the references in the previous subsection, able to be the function space of initial data $a(x)$ for which the Cauchy problem (1.1)–(1.3) admits unique time-global initial data.

We can prove Theorems 9 and 10 by rewriting the nonlinear term $(u \cdot \nabla_x)u$ as

$$\sum_{k=1}^{n} \left\{ \frac{\partial}{\partial x_k} (u_k(t,x)u_j(t,x)) \right\}_{j=1}^{n}$$

in view of (1.2), rewriting the equation into an integral equation, and solving it by successive approximation following Fujita and Kato [**27**]. Proposition 8 is used in the estimate. For details, see the original paper [**50**], or the lecture notes [**97**] in Japanese.

Furioli, Lemarié-Rieusset and Terraneo [**29**] showed the uniqueness of the solutions of (1.1)–(1.3) on $\mathbb{R}^3$ in the space $C\left([0,T), L^3\right)$, by using the Besov space $\dot{B}^{1/2}_{2,\infty}$ in an essential way. However, as we said in Remark 2, we cannot expect that the solution is strongly continuous on the interval $[0, \infty)$, since the Laplacian is not densely defined in general on our function spaces. More recently, by using the energy estimate, Lions and Masmoudi [**56**] gave another proof of the result of [**29**], and proved the uniqueness of the solution of (1.1)–(1.3) in the space $L^\infty\left([0,T], L^n\right)$ for $n \geq 4$. But in general the energy estimate cannot be applied to our problem either. Hence to find a necessary condition for the uniqueness remains an open problem. The author cannot tell so far whether the condition (2) in Theorem 9 is essential or not.

We consider other topics related to the results above. First, we can define the uniformly local Morrey spaces $M_{p,q}$ by replacing the range of R in the definition (2.4) by $0 < R \leq 1$, and define the uniformly local version of our spaces $N^s_{p,q,r}$ based on $M_{p,q}$ by using nonhomogeneous Littlewood-Paley decomposition. Then these spaces are strictly larger than $\mathcal{N}^s_{p,q,r}$ for $s < 0$, and we can show the unique existence of time-local solutions of the Cauchy problem (1.1)–(1.3) for $a(x) \in N^s_{p,q,r}$. (In fact, Theorem 9 is still valid if the space $\mathcal{N}^{n/p-1}_{p,q,\infty}$ is replaced by $N^{n/p-1}_{p,q,\infty}$.) Contrary to the case for the usual L^p-spaces, we cannot show the local existence for arbitrary $a(x) \in N^{n/p-1}_{p,q,\infty}$, but the singularity of $a(x)$, namely the distance of $a(x)$ from the closure of $\mathcal{S}$ in $N^{n/p-1}_{p,q,\infty}$, must be sufficiently small. This fact is closely related to Remark 2. As is verified in Giga, Miyakawa and Osada [**34**] and Kato [**44**], conditions of this type are not necessary for the existence of the solution in the case $n = 2$. But it is unknown so far whether they are necessary for the uniqueness in the case $n = 2$, and for both uniqueness and existence in the case $n \geq 3$.

We can consider similar problems for the Fujita-type semilinear heat equations (2.3), and Brezis and Friedman [**11**], Baras and Pierre [**2**] and Niwa [**68**] studied them with measures as initial data. Kozono and Yamazaki [**50**] gave conditions sufficient for unique time-global solvability for initial data in $\mathcal{N}^s_{p,q,\infty}$, and conditions sufficient for unique time-local solvability for initial data in $N^s_{p,q,\infty}$. These results imply the existence of initial data other than Radon measures for which the Cauchy problem is uniquely time-globally solvable. For example, put $p = n(\gamma - 1)/(3 - \gamma)$ in the case $n \geq 3$ and $(3 - n + \sqrt{n^2 - 2n + 9})/2 < \gamma < 2$. There exists a unique time-global solution of (2.3) with initial data

$$c\,\mathrm{p.f.}(x_1)_+^{-1-1/p} \cdot \prod_{j=2}^{n} (x_j)_+^{1/p}$$

provided $|c|$ is sufficiently small. Here p. f. denotes the finite part of functions that are not locally integrable.

For semilinear heat equations of Fujita type, the existence of null solutions constructed by Haraux and Weissler [**37**] implies that some sort of condition corresponding to condition (2) of Theorem 9 is essential for the uniqueness, as Wu [**96**] recently pointed out. Furthermore, Weissler [**93**] showed that there are no time-local solutions if the singularity of the initial data is large.

Suppose that $n \geq 3$, $2 < r \leq n$, and that the external force $f(x)$ is independent of t and can be written as a linear combination of second-order derivatives of functions belonging to $\mathcal{M}_{n,r}$. Then Kozono and Yamazaki [**52**] showed the unique existence of a solution $w(x) \in \mathcal{M}_{n,r}$ with sufficiently small $\| w \,|\mathcal{M}_{n,r} \|$ of the stationary problem (1.7)–(1.8), and proved the unique existence of time-global solutions, together with the decay estimates, of the equation (1.9)–(1.11). Zhou [**100**] recently solved the corresponding problem for (2.3).

In order to take a function in a space larger than L^n as initial data for the initial boundary value problem with the Dirichlet boundary condition in exterior domains, one needs the theory of the Helmholtz decomposition in the space. Miyakawa and Yamada [**66**] constructed this theory for the spaces $L^{p,\infty}$ by using real interpolation from the theory for the usual L^p-spaces given by Fujiwara and Morimoto [**28**] and Miyakawa [**62**]. Based on this theory, Kozono and Yamazaki [**51**] proved the unique existence of time-global solutions with small initial data in $L^{n,\infty}$, and that of time-local solutions with initial data in $L^{n,\infty}$ with sufficiently small singularity.

Furthermore, for $n = 2$, it has been proved that time-local solutions are necessarily prolonged into time-global solutions. Borchers and Miyakawa [**9**] and Kozono and Yamazaki [**54**] considered the case $n \geq 3$ with external force independent of the time variable, and showed the unique existence of time-global solutions and investigated their asymptotic behavior for small initial disturbances in $L^{n,\infty}$ under the assumption that the stationary solution is sufficiently small in $L^{n,\infty}$.

3. Two-dimensional Poisson equation

In this section we set $n = 2$, and state the result of Evans and Müller [**23**] on the estimate of solutions to the Poisson equation on $\mathbb{R}^2$ in the local Hardy space and its application to the two-dimensional Euler equation. First, as a local version of the maximal function introduced in Section 1, we introduce the function $\psi^* u$, defined as

$$(\psi^* u)(x) = \sup_{0<t\leq 1} \frac{1}{t^n} \left| \left(\psi\left(\frac{\cdot}{t}\right) * u \right)(x) \right| = \sup_{0<t\leq 1} \left| \int_{\mathbb{R}^n} \psi(y) u(x - ty) dy \right|$$

for $u \in \mathcal{S}'$ and $\psi \in \mathcal{S}$ such that $\int_{\mathbb{R}^n} \psi(x)dx = 1$. Then, following Goldberg [**35**], we define the local Hardy space h^1 by

$$h^1 = \left\{ u \in \mathcal{S}' \mid \left\| u \,|h^1 \right\| = \left\| \psi^* u \,|L^1 \right\| < \infty \right\}.$$

Then Evans and Müller [**23**] proved the following theorem.

THEOREM 11. *For every $\varphi \in C_0^\infty(\mathbb{R}^2)$, there exist positive constants C and R such that for every function $\omega \in L^1_{\mathrm{loc}}$ with $\omega(x) \geq 0$ for almost all $x \in \mathbb{R}^2$, we have the estimate*

$$\left\| \varphi \frac{\partial u}{\partial x_1} \frac{\partial u}{\partial x_2} \middle| h^1 \right\| + \left\| \varphi \left\{ \left(\frac{\partial u}{\partial x_1} \right)^2 - \left(\frac{\partial u}{\partial x_2} \right)^2 \right\} \middle| h^1 \right\| \leq C \int_{|x|<R} \left(\left| \frac{\partial u}{\partial x_1} \right|^2 + \left| \frac{\partial u}{\partial x_2} \right|^2 \right) dx \tag{3.1}$$

for the derivatives of the solution $u(x)$ of the Poisson equation $-\Delta u = \omega$.

By using this theorem and a compensated compactness argument, Evans and Müller [**23**] gave another proof of the result of Delort [**22**] on the existence of weak solutions to the Cauchy problem for the two-dimensional Euler equation with nonnegative measures as initial vorticity. Evans and Müller also showed that (3.1) holds under the assumption that $\omega(x)$ is rotationally symmetric with respect to the origin, and gave a counterexample of $\omega(x)$ with which (3.1) fails to hold.

Semmes [**82**] gave another proof of Theorem 11. On the other hand, Shimizu [**84**] gave another counterexample, and proved that (3.1) does not hold in general even if $\omega(x) \geq 0$ for the Poisson equation on the half plane under the Dirichlet boundary condition.

4. Unsolved problems

I intended to discuss very limited topics in detail, but the explanation, particularly the outline of proof, seems to be far from enough, because the definition of function spaces took so much space. I will close this article by listing a few unsolved problems.

Most of the theory discussed in this article is limited to the solution on the whole space $\mathbb{R}^n$, due mainly to the technical difficulty in defining function spaces on general domains. But it is natural to consider similar problems on general domains, particularly exterior domains.

A problem on the decay of weak solutions on $\mathbb{R}^n$ is whether we can replace the Hardy space H^p by the space X^p introduced by (1.4). It seems difficult to do so for general problems [**19**], but it might be possible in the problems treated in [**63, 64, 65**].

Another problem of interest on the same topic is the question whether the decay order on the half plane is the same as that on the whole plane, since the same conclusion does not hold on exterior domains, as we have seen in Section 1. Although Giga, Matsui and Shimizu [**32**] and Shimizu [**85**] gave the estimate of the Stokes semigroup in the Hardy spaces, which might be helpful to solve this problem, the original question is still open. Another question is to obtain the best decay order in exterior domains.

In order to consider the Cauchy problem with distributions other than Radon measures as initial data on general domains, it is necessary to characterize traces of the elements of $\mathcal{N}^s_{p,q,r}$, and construct the theory of the Helmholtz decomposition on this basis. Characterizing the trace would be of interest in the application of these spaces to other partial differential equations, as well as its own interest in the theory of function spaces. The question whether there is a blowing-up solution among the pseudo self-similar solutions introduced in [**59**] is interesting, and the space $\mathcal{N}^s_{p,q,r}$ might be helpful to answer it.

The problems in Sections 1 and 2 are also considered for the problem (1.9)–(1.11), and similar results are obtained in the case $n \geq 3$. However, very little is known when the external force depends on the time variable, such as perturbation of time-periodic solutions. Although unique existence and stability of solutions in the Morrey spaces have recently been proved in [**98**], the results are much less satisfactory than the results in the space $L^{n,\infty}$ obtained in [**99**]. The reason is that the Morrey spaces are known to be the dual spaces of some function spaces, but very little is known for such predual spaces. (See Long [**57**] and Kalita [**41**].)

Furthermore, as is stated in Shibata [**83**], very little is known for the two-dimensional stationary Navier-Stokes equation. Analysis on exterior domains is more difficult than on the whole spaces, since the decay of solutions is slower, due to the net force exerted on the boundary. Hence we should first limit ourselves to the problem on the whole plane, and try to find a function space on which the existence, the uniqueness and the stability of small solutions are proved simultaneously, under appropriate assumptions on the external force. But nothing has yet been obtained on this problem. The difficulty can be understood in view of the results of [**23**], particularly the fact that (3.1) does not hold in general.

Finally, the author expresses his sincere gratitude to Prof. Yoshikazu Giga of Hokkaido University for much advice, from the choice of topics to the details of expression. He is also grateful to Prof. Hideo Kozono of Tôhoku University, Prof. Takayoshi Ogawa of Kyushu University and Prof. Naoyuki Ishimura of Hitotsubashi University for pointing out errors and giving much information on the references.

Note added in proof. After completing the manuscript, the author found Koch and Tataru [**101**]. It proved the time-global well-posedness of (1.1)–(1.3) with small $a(x) \in BMO^{-1}$, the space consisting of finite linear combinations of first-order derivatives of the functions in BMO. The space BMO^{-1} contains $\dot{B}^{n/p-1}_{p,\infty}$ for $p < \infty$, $L^{n,\infty}$ and $\mathcal{M}_n$. Hence this result covers the results of [**3, 4, 14, 15, 16, 33, 43, 74, 75, 87, 92**]. However, the author cannot tell so far that this result surely covers the result of [**50**], although it is very likely.

References

[1] Arai, H., *Advance of real analysis and its influences to analysis* (in Japanese), Sūgaku **50** (1998), 29–55.

[2] Baras, P. and Pierre, M., *Problèmes paraboliques semi-linéaires avec données mesures*, Applicable Anal., **18** (1984), 111–149.

[3] Barraza, O. A., *Self-similar solutions in weak L^p-spaces of the Navier-Stokes equations*, Rev. Mat. Iberoamericana, **12** (1996), 411–439.

[4] Barraza, O. A., *Regularity and stability for the solutions of the Navier-Stokes equations in Lorentz spaces*, Nonlinear Anal., **35** (1999), 747–764.

[5] Beale, J. T., Kato, T. and Majda, A., *Remarks on the breakdown of smooth solutions for the 3-D Euler equations*, Comm. Math. Phys., **94** (1984), 61–66.

[6] Bergh, J. and Löfström, J., *Interpolation Spaces. An Introduction*, Grundlehren der Mathematischen Wissenschaften **223**, Springer, Berlin, 1976.

[7] Borchers, W. and Miyakawa, T., *Algebraic L^2 decay for Navier-Stokes flows in exterior domains*, Acta Math., **165** (1990), 189–227.

[8] Borchers, W. and Miyakawa, T., *Algebraic L^2 decay for Navier-Stokes flows in exterior domains*, II, Hiroshima Math. J., **21** (1991), 621–640.

[9] Borchers, W. and Miyakawa, T., *On stability of exterior stationary Navier-Stokes flows*, Acta Math., **174** (1995), 311–382.

[10] Bourdaud, G., *Réalisations des espaces de Besov homogènes*, Ark. Mat., **26** (1988), 41–54.

[11] Brézis, H. and Friedman, A., *Nonlinear parabolic equations involving measures as initial conditions*, J. Math. Pures Appl. (9), **62** (1983), 73–97.

[12] Calderón, A. P. and Zygmund, A., *Singular integral operators and differential equations*, Amer. J. Math., **79** (1957), 901–921.

[13] Campanato, S., *Proprietà di una famiglia di spazi funzionali*, Ann. Scuola Norm. Sup. Pisa, **18** (1964), 137–160.

[14] Cannone, M., *Ondelettes, paraproduits et Navier-Stokes*, Diderot, Paris, 1995.

[15] Cannone, M., *A generalization of a theorem by Kato on Navier-Stokes equations*, Rev. Mat. Iberoamericana, **13** (1997), 515–541.

[16] Cannone, M. and Planchon, F., *Self-similar solutions for Navier-Stokes equations in $\mathbb{R}^3$*, Comm. Partial Differential Equations, **21** (1996), 179–193.

[17] Carpio, A., *Asymptotic behavior for the vorticity equations in dimension two and three*, Comm. Partial Differential Equations, **19** (1994), 827–872.

[18] Carpio, A., *Large-time behavior in incompressible Navier-Stokes equations*, SIAM J. Math. Anal., **27** (1996), 449–475.

[19] Coifman, R. R., Lions, P.-L., Meyer, Y. and Semmes, S., *Compensated compactness and Hardy spaces*, J. Math. Pures Appl. (9), **72** (1993), 247–286.

[20] Cottet, G.-H., *Équations de Navier-Stokes dans le plan avec tourbillon initial mesure*, C. R. Acad. Sci. Paris, Sér. I, **303** (1986), 105–108.

[21] Cottet, G.-H. and Soler, J., *Three-dimensional Navier-Stokes equations for singular filament initial data*, J. Differential Equations, **74** (1988), 234–253.

[22] Delort, J.-M., *Existence de nappes de tourbillon en dimension deux*, J. Amer. Math. Soc., **4** (1991), 553–586.

[23] Evans, L. C. and Müller, S., *Hardy spaces and the two-dimensional Euler equations with nonnegative vorticity*, J. Amer. Math. Soc., **7** (1994), 199–219.

[24] Federbush, P., *Navier and Stokes meet the wavelet*, Comm. Math. Phys., **155** (1993), 219–248.

[25] Fefferman, C. and Stein, E. M., *H^p spaces of several variables*, Acta Math., **129** (1972), 137–193.

[26] Fujita, H., *On the blowing up of solutions of the Cauchy problem for $u_t = \Delta u + u^{1+\alpha}$*, J. Fac. Sci. Univ. Tokyo, Sec. I, **13** (1966), 109–124.

[27] Fujita, H. and Kato, T., *On the Navier-Stokes initial value problem I*, Arch. Rational Mech. Anal., **16** (1964), 269–315.

[28] Fujiwara, D. and Morimoto, H., *An L^r-theorem of the Helmholtz decomposition of vector fields*, J. Fac. Sci. Univ. Tokyo, Sec. IA, **24** (1977), 685–700.

[29] Furioli, G., Lemarié-Rieusset, P. G. and Terraneo, E., *Sur l'unicité dans $L^3(\mathbb{R}^3)$ des solutions mild des équations de Navier-Stokes*, C. R. Acad. Sci. Paris, Sér. I., **325** (1997), 1253–1256.

[30] Giga, Y. and Giga, M.-H., *Nonlinear Partial Differential Equations—Asymptotic Behavior of Solutions and Self-Similar Solutions* (in Japanese), Kyôritsu Shuppan, Tokyo, 1999.

[31] Giga, Y. and Kambe, T., *Large time behaviour of the vorticity of two-dimensional viscous flow and its application to vortex formation*, Comm. Math. Phys., **117** (1988), 549–568.

[32] Giga, Y., Matsui, S. and Shimizu, Y., *On estimates in Hardy spaces for the Stokes flow in a half space*, Math. Z., **231** (1999), 383–396.

[33] Giga, Y. and Miyakawa, T., *Navier-Stokes flow in $\mathbb{R}^3$ with measures as initial vorticity and Morrey spaces*, Comm. Partial Differential Equations, **14** (1988), 577–618.

[34] Giga, Y., Miyakawa, T. and Osada, H., *Two-dimensional Navier-Stokes flow with measures as initial vorticity*, Arch. Rational Mech. Anal., **104** (1988), 233–250.

[35] Goldberg, D., *A local version of real Hardy spaces*, Duke Math. J., **46** (1979), 27–42.

[36] Grubb, G., *Initial value problems for the Navier-Stokes equations with Neumann boundary conditions*, The Navier-Stokes Equations II, J. G. Heywood, K. Masuda, R. Rautmann and S. A. Solonnikov, eds., Proc. Conf. Oberwolfach 1991, Lecture Notes in Math., **1530**, Springer, Berlin, 1992, pp. 262–283.

[37] Haraux, A. and Weissler, F., *Non-uniqueness for a semilinear initial value problem*, Indiana Univ. Math. J., **31** (1982), 167–189.

[38] Hishida, T., *On a class of stable steady flows to the exterior convection problem*, J. Differential Equations, **141** (1997), 54–85.

[39] Jawerth, B., *Some observations on Besov and Triebel-Lizorkin spaces*, Math. Scand., **40** (1977), 94–104.

[40] Kajikiya, R. and Miyakawa, T., *On L^2 decay of weak solutions of the Navier-Stokes equations in $\mathbb{R}^n$*, Math. Z., **192** (1986), 135–148.

[41] Kalita, E. A., *Dual Morrey spaces*, Doklady Akad. Nauk, **361** (1998), 447–449; Engl. transl., Doklady Math., **58** (1998), 85–87.

[42] Kato, T., *Strong L^p-solutions of the Navier-Stokes equation in $\mathbb{R}^n$, with applications to weak solutions*, Math. Z., **187** (1984), 471–480.

[43] Kato, T., *Strong solutions of the Navier-Stokes equations in Morrey spaces*, Boll. Soc. Brasil. Mat. (N. S.), **22** (1992), 127–155.

[44] Kato, T., *The Navier-Stokes equation for an incompressible fluid in $\mathbb{R}^2$ with a measure as initial vorticity*, Differential and Integral Equations, **7** (1994), 949–966.

[45] Kato, T. and Ponce, G., *The Navier-Stokes equation with weak initial data*, Internat. Math. Res. Notices, **1994**, 435–444.

[46] Kobayashi, T. and Muramatu, T., *Abstract Besov space approach to the non-stationary Navier-Stokes equations*, Math. Methods Appl. Sci., **15** (1992), 599–620.

[47] Kozono, H., *L^1-solutions of the Navier-Stokes equations in exterior domains*, Math. Ann., **312** (1998), 319–340.

[48] Kozono, H. and Taniuchi, Y., *Bilinear estimates in BMO and the Navier-Stokes equations*, Math. Z., **235** (2000), 173–194.

[49] Kozono, H. and Taniuchi, Y., *Limiting case of the Sobolev inequality and BMO, with applications to the Euler equations*, preprint.

[50] Kozono, H. and Yamazaki, M., *Semilinear heat equations and the Navier-Stokes equation with distributions in new function spaces as initial data*, Comm. Partial Differential Equations, **19** (1994), 959–1014.

[51] Kozono, H. and Yamazaki, M., *Local and global unique solvability of the Navier-Stokes exterior problem with Cauchy data in the space $L^{n,\infty}$*, Houston J. Math., **21** (1995), 755–799.

[52] Kozono, H. and Yamazaki, M., *The stability of small stationary solutions in Morrey spaces of the Navier-Stokes equation*, Indiana Univ. Math. J., **44** (1995), 1307–1336.

[53] Kozono, H. and Yamazaki, M., *Exterior problem for the stationary Navier-Stokes equations in the Lorentz spaces*, Math. Ann., **310** (1998), 279–305.

[54] Kozono, H. and Yamazaki, M., *On a larger class of stable solutions to the Navier-Stokes equations in exterior domains*, Math. Z., **228** (1998), 751–785.

[55] Leray, J., *Sur le mouvement d'un liquide visqueux emplissant l'espace*, Acta Math., **63** (1934), 193–248.

[56] Lions, P.-L. et Masmoudi, N., *Unicité des solutions faibles de Navier-Stokes dans $L^N(\Omega)$*, C. R. Acad. Sci. Paris, Sér. I, **327** (1998), 491–496.

[57] Long, R., *The spaces generated by blocks*, Scientia Sinica, Ser. A, **27** (1984), 16–26.

[58] Majda, A., *Vorticity and the mathematical theory of incompressible fluid flow*, Comm. Pure Appl. Math., **39** (1986), S187–S220.

[59] Márek, J., Nečas, J., Pokorný, M. and Schonbek, M. E., *On possible singular solutions to the Navier-Stokes equations*, Math. Nachr., **199** (1999), 97–114.

[60] Masuda, K., *Weak solutions of Navier-Stokes equations*, Tôhoku Math. J. (2), **36** (1984), 623–646.

[61] Michaux, M. and Rakotoson, J. M., *Remarks on Navier-Stokes equations with measures as data*, Appl. Math. Lett., **6-6** (1993), 75–77.

[62] Miyakawa, T., *On nonstationary solutions of the Navier-Stokes equations in an exterior domain*, Hiroshima Math. J., **12** (1982), 115–140.

[63] Miyakawa, T., *Hardy spaces of solenoidal vector fields, with applications to the Navier-Stokes equations*, Kyushu J. Math., **50** (1996), 1–64.

[64] Miyakawa, T., *On L^1-stability of stationary Navier-Stokes flows in $\mathbb{R}^n$*, J. Math. Sci. Univ. Tokyo, **4** (1997), 67–119.

[65] Miyakawa, T., *Application of Hardy space techniques to the time-decay problem for incompressible Navier-Stokes flows in $\mathbb{R}^n$*, Funkcialaj Ekvacioj, **41** (1998), 383–434.

[66] Miyakawa, T. and Yamada, M., *Planar Navier-Stokes flows in a bounded domain with measures as initial vorticities*, Hiroshima Math. J., **22** (1992), 401–420.

[67] Nečas, J., Růžička, M. and Šverák, V., *On Leray's self-similar solutions of the Navier-Stokes equations*, Acta Math., **176** (1996), 283–294.

[68] Niwa, Y., *Semilinear heat equations with measures as initial data*, Thesis, University of Tokyo, 1986.

[69] Novotny, A. and Padula, M., *Note on decay of solutions of steady Navier-Stokes equations in 3-D exterior domains*, Differential and Integral Equations, **8** (1995), 1833–1842.

[70] Okamoto, H., *A uniqueness theorem for unbounded classical solution of the nonstationary Navier-Stokes equations in $\mathbb{R}^3$*, J. Math. Anal. Appl., **181** (1994), 473–482.

[71] Okamoto, H., *Exact solutions of the Navier-Stokes equations via Leray's scheme*, Japan J. Indust. Appl. Math., **14** (1997), 169–197.

[72] Ozawa, T., *Scattering theory for nonlinear Schrödinger equations* (in Japanese), Sūgaku, **50** (1998), 337–349.

[73] Peetre, J., *On the theory of $L_{p,\lambda}$ spaces*, J. Funct. Anal., **4** (1969), 71–87.

[74] Planchon, F., *Global strong solutions in Sobolev or Lebesgue spaces to the incompressible Navier-Stokes equations in $\mathbb{R}^3$*, Ann. Inst. Henri Poincaré, Anal. Non Linéaire, **13** (1996), 319–336.

[75] Planchon, F., *Asymptotic behavior of global solutions to the Navier-Stokes equations in $\mathbb{R}^3$*, Rev. Mat. Iberoamericana, **14** (1998), 71–93.

[76] Ponce, G., *Remarks on a paper by J. T. Beale, T. Kato and A. Majda*, Comm. Math. Phys., **98** (1985), 349–353.

[77] Schonbek, M. E., *L^2 decay for weak solutions of the Navier-Stokes equations*, Arch. Rational Mech. Anal., **88** (1985), 209–222.

[78] Schonbek, M. E., *Large time behaviour of solutions to the Navier-Stokes equations*, Comm. Partial Differential Equations, **11** (1986), 733–763.

[79] Schonbek, M. E., *Lower bounds of rates of decay for solutions to the Navier-Stokes equations*, J. Amer. Math. Soc., **4** (1991), 423–449.

[80] Schonbek, M. E., *Asymptotic behavior of solutions to the three-dimensional Navier-Stokes equations*, Indiana Univ. Math. J., **41** (1992), 809–823.

[81] Schonbek, M. E., Schonbek, T. P. and Süli, E., *Large-time behaviour of solutions to the magneto-hydrodynamics equations*, Math. Ann., **304** (1996), 717–756.

[82] Semmes, S., *A primer on Hardy spaces, and some remarks on a theorem of Evans and Müller*, Comm. Partial Differential Equations, **19** (1994), 277–319.

[83] Shibata, Y., *Global solutions of nonlinear evolution equations and their stability* (in Japanese), Sūgaku, **51** (1999), 1–17.

[84] Shimizu, Y., *A remark on estimates of bilinear forms of gradients in Hardy space*, Proc. Japan Acad., Ser. A, **72** (1996), 179–182.

[85] Shimizu, Y., *L^∞-estimates of first-order space derivatives of Stokes flow in a half space*, Funkcialaj Ekvacioj, **42** (1999), 291–309.

[86] Stein, E. M., *Harmonic Analysis: Real-Variable Methods, Orthogonality, and Oscillatory Integrals*, Princeton Mathematical Series, Vol. 43, Princeton University Press, Princeton, 1993.

[87] Taylor, M. E., *Analysis on Morrey spaces and applications to Navier-Stokes and other evolution equations*, Comm. Partial Differential Equations, **17** (1992), 1407–1456.

[88] Tian, G. and Xin, Z., *One-point singular solutions to the Navier-Stokes equations*, Topological Methods in Nonlinear Anal., **11** (1998), 135–145.

[89] Triebel, H., *Theory of Function Spaces*, Monographs in Mathematics, Vol. 78, Birkhäuser, Basel, 1983.

[90] Triebel, H., *Theory of Function Spaces* II, Monographs in Mathematics, Vol. 84, Birkhäuser, Basel, 1992.

[91] Tsai, T.-P., *On Leray's self-similar solutions of the Navier-Stokes equations satisfying local energy estimates*, Arch. Rational Mech. Anal., **143** (1998), 29–51; Erratum, ibid., **147** (1999), 363.

[92] Vishik, M., *Hydrodynamics in Besov spaces*, Arch. Rational Mech. Anal., **145** (1998), 197–214.

[93] Weissler, F., *Local existence and nonexistence for semilinear parabolic equations in L^p*, Indiana Univ. Math. J., **29** (1980), 79–102.

[94] Wiegner, M., *Decay results for weak solutions of the Navier-Stokes equations in $\mathbb{R}^n$*, J. London Math. Soc., **35** (1987), 303–313.

[95] Wiegner, M., *The Navier-Stokes equations—A neverending challenge?*, Jahresber. Deutsch. Math.-Verein., **101** (1999), 1–25.

[96] Wu, J., *Well-posedness of a semilinear heat equation with weak initial data*, J. Fourier Anal. Appl., **4** (1998), 629–642.

[97] Yamazaki, M., *Solutions of the Navier-Stokes equations with distributions as initial data* (in Japanese), Graduate School of Mathematical Sciences, University of Tokyo, Tokyo, 1999.

[98] Yamazaki, M., *Solutions in the Morrey spaces of the Navier-Stokes equation with time-dependent external force*, Funkcialaj Ekvacioj, **43** (2000), 419–460.

[99] Yamazaki, M., *The Navier-Stokes equations in the weak-L^n space with time-dependent external force*, Math. Ann., **317** (2000), 635–675.

[100] Zhou, X., *The stability of small stationary solutions in Morrey spaces of the semilinear heat equations*, J. Math. Sci. Univ. Tokyo, **6** (1999), 793–822.

[101] Koch, H. and Tataru, D., *Well-posedness for the Navier-Stokes equations*, Adv. Math., **157** (2001), 22–35.

Department of Mathematics, Graduate School of Economics, Hitotsubashi University, Kunitachi, Tokyo 186-8601 Japan

E-mail address: `cc00053@srv.cc.hit-u.ac.jp`

Translated by MASAO YAMAZAKI

Selected Titles in This Series

(Continued from the front of this publication)

175 **R. L. Dobrushin, R. A. Minlos, M. A. Shubin, and A. M. Vershik, Editors,** Contemporary Mathematical Physics (F. A. Berezin Memorial Volume)

174 **A. A. Bolibruch, A. S. Merkur′ev, and N. Yu. Netsvetaev, Editors,** Mathematics in St. Petersburg

173 **V. Kharlamov, A. Korchagin, G. Polotovskiĭ, and O. Viro, Editors,** Topology of Real Algebraic Varieties and Related Topics

172 **K. Nomizu, Editor,** Selected Papers on Number Theory and Algebraic Geometry

171 **L. A. Bunimovich, B. M. Gurevich, and Ya. B. Pesin, Editors,** Sinai's Moscow Seminar on Dynamical Systems

170 **S. P. Novikov, Editor,** Topics in Topology and Mathematical Physics

169 **S. G. Gindikin and E. B. Vinberg, Editors,** Lie Groups and Lie Algebras: E. B. Dynkin's Seminar

168 **V. V. Kozlov, Editor,** Dynamical Systems in Classical Mechanics

167 **V. V. Lychagin, Editor,** The Interplay between Differential Geometry and Differential Equations

166 **O. A. Ladyzhenskaya, Editor,** Proceedings of the St. Petersburg Mathematical Society, Volume III

165 **Yu. Ilyashenko and S. Yakovenko, Editors,** Concerning the Hilbert 16th Problem

164 **N. N. Uraltseva, Editor,** Nonlinear Evolution Equations

163 **L. A. Bokut′, M. Hazewinkel, and Yu. G. Reshetnyak, Editors,** Third Siberian School "Algebra and Analysis"

162 **S. G. Gindikin, Editor,** Applied Problems of Radon Transform

161 **K. Nomizu, Editor,** Selected Papers on Analysis, Probability, and Statistics

160 **K. Nomizu, Editor,** Selected Papers on Number Theory, Algebraic Geometry, and Differential Geometry

159 **O. A. Ladyzhenskaya, Editor,** Proceedings of the St. Petersburg Mathematical Society, Volume II

158 **A. K. Kelmans, Editor,** Selected Topics in Discrete Mathematics: Proceedings of the Moscow Discrete Mathematics Seminar 1972–1990

157 **M. Sh. Birman, Editor,** Wave Propagation. Scattering Theory

156 **V. N. Gerasimov, N. G. Nesterenko, and A. I. Valitskas,** Three Papers on Algebras and Their Representations

155 **O. A. Ladyzhenskaya and A. M. Vershik, Editors,** Proceedings of the St. Petersburg Mathematical Society, Volume I

154 **V. A. Artamonov et al.,** Selected Papers in K-Theory

153 **S. G. Gindikin, Editor,** Singularity Theory and Some Problems of Functional Analysis

152 **H. Draškovičová et al.,** Ordered Sets and Lattices II

151 **I. A. Aleksandrov, L. A. Bokut′, and Yu. G. Reshetnyak, Editors,** Second Siberian Winter School "Algebra and Analysis"

150 **S. G. Gindikin, Editor,** Spectral Theory of Operators

149 **V. S. Afraĭmovich et al.,** Thirteen Papers in Algebra, Functional Analysis, Topology, and Probability, Translated from the Russian

148 **A. D. Aleksandrov, O. V. Belegradek, L. A. Bokut′, and Yu. L. Ershov, Editors,** First Siberian Winter School "Algebra and Analysis"

147 **I. G. Bashmakova et al.,** Nine Papers from the International Congress of Mathematicians, 1986

For a complete list of titles in this series, visit the AMS Bookstore at **www.ams.org/bookstore/**.